TOMORROW'S WORLD: **SPACE TECHNOLOGY**

USA
NASA
Challenger

TOMORROW'S WORLD

# Space Technology

Max Whitby

BBC Publications

For CFC



**Acknowledgments**

The research for several of the following chapters was undertaken during the making of the Horizon documentary *Beyond The Moon* which, like this book, examines the next ten years in space. I would particularly like to thank Graham Massey, Editor of Horizon, and Andrew Mitchell, Assistant Producer on the programme. I am also grateful to many kind and helpful people from NASA Public Affairs, including Don Bane at JPL, Jim Elliot at Goddard, Ed Harrison at KSC, Bill O'Donnell in Washington, Tim Tyson at MSFC and Sharon Wanglin at the Space Telescope Science Institute. Finally I would like to thank the pirates at Cupertino and the proprietress of Sandpond Cottage.

*Tomorrow's World* was first broadcast on British television in 1965 and has been screened every year since then.

Published by BBC Publications
a division of BBC Enterprises Ltd,
35 Marylebone High Street, London W1M 4AA

This book was designed and produced by
The Oregon Press Limited, Faraday House,
8 Charing Cross Road, London WC2H 0HG

First published 1986


ISBN 0 563 20380 3 hardback
ISBN 0 563 20378 1 paperback

Design: Martin Bristow
Picture Research: Susan Bolsom-Morris
Reader: Raymond Kaye

Illustrations: MJL Cartographics

Filmset by SX Composing Ltd, Rayleigh, England
Printed and bound in Spain by Printer Industria Gràfica SA,
Barcelona D.L.B. 4134-1986

FRONTISPIECE Launch crews work throughout the night to ready the ill-fated Challenger for an early morning departure. NASA's three remaining operational Shuttles are the foundation of the United States' space programme.

TITLE PAGE Galileo, the most advanced robot spacecraft ever sent to explore the solar system, is depicted skimming above the north pole of the Jovian moon Europa.

# Contents

USA
United States

CHAPTER ONE

# The Shuttle

*Obviously a major malfunction . . .*

Stephen Nesbitt, the voice of Mission Control

*Challenger* and her crew of seven astronauts disintegrated in a massive fireball nine miles above the Florida coastline late in the morning of Tuesday 28 January 1986. As the debris rained towards the ocean, the Shuttle's two solid rocket boosters (one of them the probable cause of the catastrophe) emerged from the flames and snaked an erratic course across the sky. It looked for all the world like a sign from the devil: the twin rocket trails tracing the shape of a twisted pair of horns on the head of the exploding spacecraft.

The loss of *Challenger* is an event from which the American space programme will take years to recover. But the tragedy reminds us how much space exploration still matters. Across the planet people were in tears, mourning the death of five men and two women who symbolise our aspirations and our frailty.

## Fire in the tail

Before *Challenger* exploded, the claim was often heard that Shuttle operations had become routine. Mission 51-L was the 25th flight of the Shuttle and the American public had become so blasé that only one cable TV network carried live pictures of the launch. No-one at NASA shared this complacency. Cold calculations had indicated that there was a 50 per cent probability of losing a Shuttle once in every hundred missions: in the event the wheels of fate turned faster than expected. Nothing has ever been 'routine' at Kennedy Space Centre when a Shuttle is about to fly.

Triumph . . . *Atlantis*, NASA's fourth operational Space Shuttle, rises magnificently from the launch pad on her maiden flight.

Disaster . . . *Challenger* disintegrates in a massive explosion 74 seconds after lift-off on 28 January 1986.

A launch is an awesome thing to witness: a controlled explosion of orange fire that catapults two million kilogrammes of high-precision machinery straight up into the sky and out of sight. The Shuttle roars away from its pad like a backwards bolt of lightning. In moments it is gone, and if everything goes well, all that remains is distant thunder and a thin vertical column of smoke twisting upwards into space.

As the countdown approaches the zero mark, designated 'T' in Nasa-speak, events begin to occur in quick succession. At T minus 3.8 seconds, the Shuttle's on-board computers instruct the first main engine to ignite, followed at 120 millisecond intervals by the second engine and then the third. These three Space Shuttle main engines (SSMEs) burn liquid oxygen and liquid hydrogen supplied from the external tank (ET) at a rate of more than 4,000 litres per second. Development of the throttleable SSMEs has pushed cryogenic rocket technology to new limits: particularly in meeting the requirement that the re-useable engines should have a working lifetime in excess of 7 hours. Reliability problems with the SSMEs have dogged Shuttle operations and are partly to blame for cost over-runs in the programme.

Within seconds, all three engines are producing 90 per cent thrust and the entire assembly lurches forward about 1 metre under the strain. The on-board computers quickly check that key systems are performing within acceptable limits and, provided all is well, they issue the critical command to fire the solid rocket boosters (SRBs). It is at this point, since there is no way to douse the SRBs once they have ignited, that the Shuttle becomes committed to the launch.

At approximately T plus 3 seconds, when the level of thrust consistently exceeds the weight of the vehicle, explosive restraining bolts are detonated and the Shuttle rises rapidly from the pad. During the ascent, computers constantly check to see that the Shuttle is stable and rising on the desired trajectory. All five engines (the three SSMEs and the two SRBs) are independently manoeuvrable and errors can be corrected by small deflections of their rocket nozzles. The same mechanism is used to roll the Shuttle through 120 degrees about 5 seconds after lift-off, placing the Orbiter beneath the external tank during the climb. As a result of this programmed manoeuvre, the astronauts complete the fiery ride to orbit head down, suspended in their seats like bats out of hell.

It was at this stage, 74 seconds into the mission and at a height of 15 kilometres (approximately 50,000 feet), that *Challenger's* large external tank exploded with the force of a small nuclear warhead. It would appear that a failed seam in one of the solid rocket boosters allowed flames to leak out and set fire to the external tank. The Shuttle is not equipped with ejector seats, but even if it were, it is doubtful whether the crew could have survived. The explosion was so massive and so sudden there was not even time to alert the ground that anything was wrong.

In a normal flight the next important event occurs at T plus 2 minutes

Their supply of fuel exhausted, the Shuttle's twin solid rocket motors separate from the main spacecraft and parachute into the ocean.

12 seconds when all the propellant in the SRBs has been consumed. Just 130 seconds after lift-off the Shuttle has already reached an altitude of 45 km (28 miles) and a velocity more than three times the speed of sound. The two expended boosters separate from the Shuttle and begin a long arcing drop into the ocean where they are collected by special recovery ships. In due course the SRBs are refurbished with fresh propellant and fly again on up to 20 missions.

Meanwhile the Shuttle's main engines continue to burn for 6 more minutes, carrying the Orbiter on up to an altitude of 130 km (81 miles). Then, somewhat surprisingly, the Shuttle performs a power dive, descending to about 116 km (72 miles) in order to jettison the almost empty external tank. The huge cylinder breaks up as it re-enters the atmosphere and its debris falls into the Indian Ocean on the opposite side of the world to where the mission began. The ET is the only major component of the Shuttle not to be re-used. A couple more rocket burns are all that is required to insert the Shuttle into its proper orbit, which can range in altitude between 185 and 1,110 km (115 and 690 miles) – depending on such factors as mission requirements, payload weight and launch angle.

*Columbia, Discovery* and *Atlantis*, NASA's three remaining Shuttle Orbiters, are the most remarkable flying machines ever built. Each manages to be three different things at once: a powerful rocket capable of producing more than seven hundred metric tons of thrust; a versatile and sophisticated spacecraft that can deliver and collect multiple payloads into and from orbit; and a winged aeroplane that can re-enter the Earth's atmosphere, gliding half-way round the planet before landing on a conventional runway. Furthermore, each Orbiter has been designed to repeat the performance on more than 100 occasions. The Shuttle has become the workhorse of the American manned space programme and, despite the tragic loss of *Challenger*, it has revolutionised our ability to work in space.

### Life in orbit

The Shuttle was originally designed to stay in space for anything up to 30 days, although most current flights last only about a week. Usually each mission has multiple objectives, typically including delivery of one or more satellites into orbit. (The range of Shuttle operations and the main features of the Orbiter are described in more detail below.) There is room on board for a crew of up to seven people: the commander, his co-pilot, and a mix of up to five mission and payload specialists who are trained to carry out duties specific to each flight. In emergencies, for example when rescuing the crew of another Orbiter that has become disabled, a total of 10 astronauts can be squeezed in.

But even with a normal complement of four to seven people, life on board the Shuttle is definitely crowded. Crew accommodation is restricted to a single pressurized compartment at the front of the Orbiter,

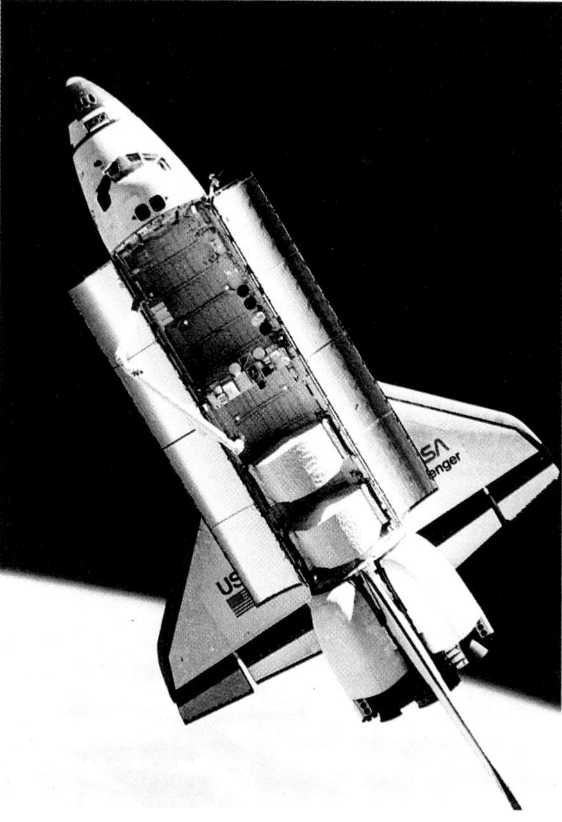

Three sides of the Shuttle: a rocket, a spacecraft and a glider.

Anatomy of the Shuttle. The European-built Spacelab system, which fits into the Shuttle's payload bay, is shown to the same scale.

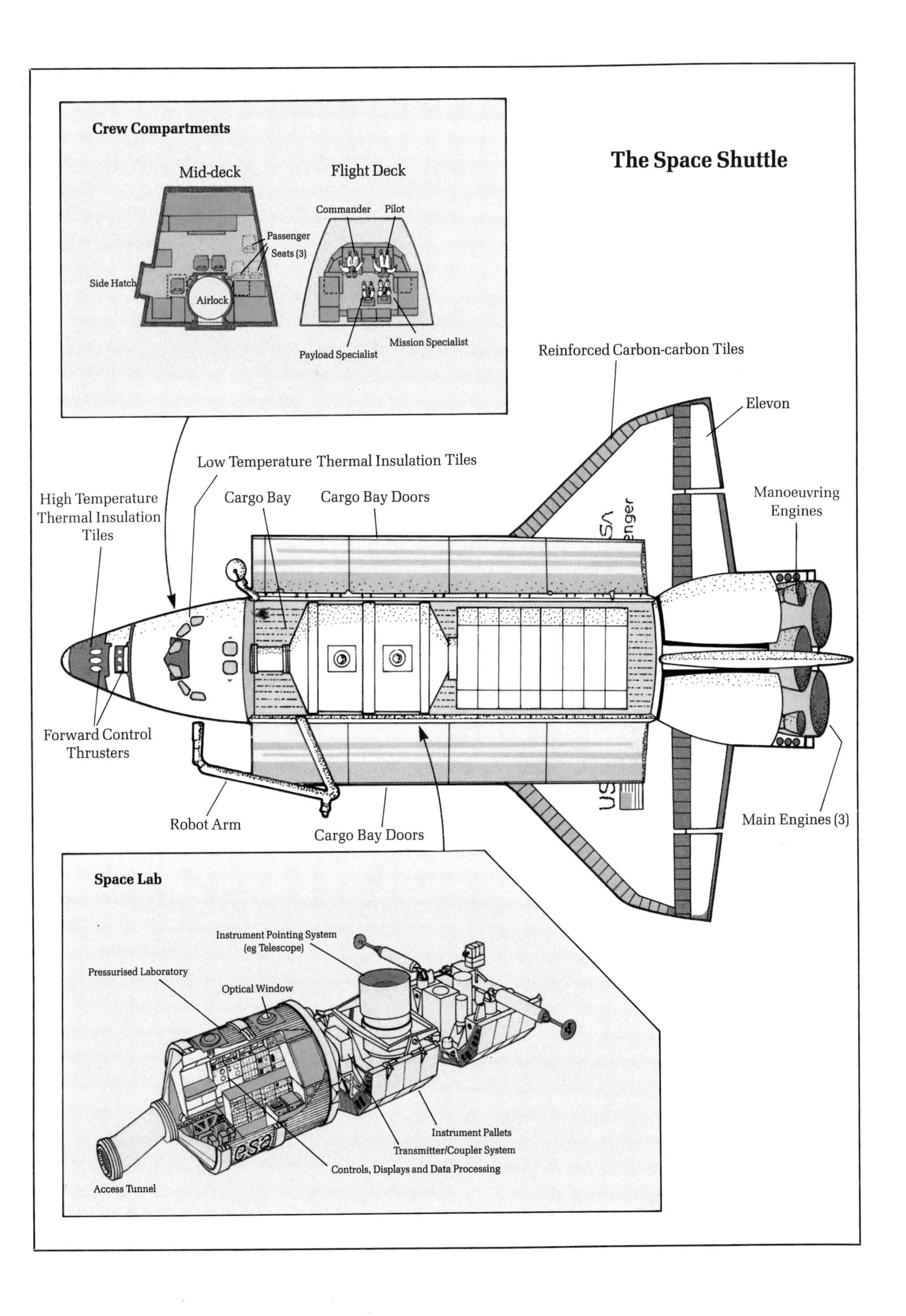
The Space Shuttle
Crew Compartments
Mid-deck
Flight Deck
Commander
Pilot
Passenger
Seats (3)
Side Hatch
Airlock
Payload Specialist
Mission Specialist
Reinforced Carbon-carbon Tiles
Elevon
Low Temperature Thermal Insulation Tiles
High Temperature
Thermal Insulation
Tiles
Cargo Bay
Cargo Bay Doors
Manoeuvring
Engines
Forward Control
Thrusters
Robot Arm
Cargo Bay Doors
Main Engines (3)
Space Lab
Instrument Pointing System
(eg Telescope)
Pressurised Laboratory
Optical Window
Instrument Pallets
Transmitter/Coupler System
Controls, Displays and Data Processing
Access Tunnel
esa

The Shuttle is covered in a mosaic of ceramic tiles that protect the Orbiter from the fierce heat of re-entry. Problems with these thermal protection tiles have plagued the Shuttle's development and operation.

which is divided into three levels. At the top is the flight deck where the commander and the pilot sit during launch and landing, looking forward out of the Shuttle's cockpit windows. Behind them are seats for two mission or payload specialists, plus various work stations that become operational in orbit.

Immediately below the flight deck is a larger area known as the mid-deck, with about 10 $m^2$ (1000 $ft^2$) of floor space and sleeping berths for four crew members. The mid-deck also includes an airlock, a galley, a 'waste management station', and room for three more seats during launch and re-entry. There is also a hatch that is used for entering the Shuttle on the pad and leaving it at the end of the mission. Finally, at the lowest level, is an area known as the equipment bay which contains most of the life-support systems and lockers for use by the crew.

When the time comes to end a mission the OMS engines fire once more, slightly slowing the Orbiter and causing it gradually to lose altitude. Still travelling at more than 26,000 km/h (16,000 mph), the Shuttle uses the outer layer of the atmosphere as an air brake, and in doing so its exterior surface begins to heat up. As re-entry proceeds, the spacecraft penetrates denser layers of the atmosphere and its temperature rises until critical areas, like the nosecap and the leading edges of the wing, may become as hot as 1430°C.

To protect the Orbiter against this fierce heat, its surface is covered with unique insulating materials that represent a breakthrough in spacecraft technology. Four types of thermal protection are used, depending on the degree of heating experienced by different regions of the airframe. The underside is covered with the infamous black ceramic tiles that caused long delays during the Shuttle's development when they repeatedly fell off during trials. And in early 1985 a new adhesive problem with the same tiles caused the first all-military Shuttle mission to be postponed for six weeks: a move that did not endear NASA to a US Air Force already disenchanted by delays and cancellations. But despite their unfortunate history, the Shuttle's thermal protection tiles deserve to be considered a success. They do their job of protecting the

Orbiter with complete efficiency, and so far, relatively few tiles have required replacement between flights.

The final event in this brief summary of a typical Shuttle mission is the landing. It is strange to think how little more than a decade ago, returning Apollo and Skylab astronauts used to splash down in the Pacific Ocean, tying up a significant part of the United States Navy in the bargain! Today the Shuttle lands just like a commercial passenger aircraft on a more-or-less ordinary runway with the minimum of fuss. Gliding without power there is only one chance to get the landing right: to save weight the undercarriage cannot be retracted once it has been lowered and, in any case, if the Shuttle overshoots the runway it will simply stall and crash.

**Delivering the goods**

The Shuttle's primary purpose in life is to deliver satellites to orbit. To this end it is equipped with a large cargo bay 18.3 m (60 ft) long and 4.6 m (15 ft) wide. The payload bay occupies the central section of the fuselage and during launch and re-entry it is enclosed by two pairs of large doors. Once in orbit, these payload bay doors swing open. Radiators are mounted on their interior surface to dissipate excess heat generated by the Shuttle's electrical systems. It is interesting to note that the large size of the payload bay, which is a critical parameter in the Orbiter's design, is dictated by an Air Force requirement to launch large military reconnaissance satellites.

The Shuttle can carry a maximum payload of 29,500 kg (65,000 lb) to low Earth orbit. (Certain launches require more energy, however, and for some missions this capacity is reduced by as much as 50 per cent.) There is room in the payload bay for up to four typically sized satellites, which are usually installed prepacked in protective cocoons that open like clams in orbit to reveal their contents. Satellites can be deployed in one of several different ways, depending on their design and the orbit they are intended to occupy.

Communications satellites, which are the Shuttle's usual commercial cargo, need to be placed in geostationary orbit at an altitude of about 35,400 km (22,000 miles). This is way above the Orbiter's operational ceiling of 1110 km (690 miles), and so additional boosters must be used to carry satellites up from the low orbit that the Shuttle *can* reach. These small rocket motors are installed beneath each satellite and the combination is released into space as a single unit. The exact method of deployment can vary, but often an electric turntable sets the satellite spinning like a top to provide stability, and then a powerful spring is released to send the revolving spacecraft on its way. Once at a safe distance from the Shuttle the rocket motor fires and, all being well, the satellite ascends to its final orbit.

Unfortunately things have not always gone well in the past. So far, the Shuttle has experienced satellite launch failures on no fewer than five

SBS-4
HUGHES

Three ways to deploy a satellite: by robot arm (top left), by 'frisbee' spin (top right), and from a spring-loaded canister (bottom).

occasions. Usually the rocket motors intended to boost the satellite to a higher orbit have been at fault. Although the Shuttle itself has worked flawlessly, these failures have naturally cast doubt on the reliability of the overall delivery system; and to make matters worse, both main types of upper stage boosters for use on the Shuttle have been involved.

The first failure was in an inertial upper stage (IUS) manufactured by Boeing Aerospace, which misfired in April 1983, taking the tracking and data relay satellite it was carrying into an unacceptably low orbit. This TDRS (pronounced 'tea dress') is a vital element in NASA's space communications network, allowing high-volume uninterrupted data exchange between mission control and numerous orbiting spacecraft including the Shuttle itself. Fortunately, ground controllers were able to use the small stabilization thrusters of the TDRS gradually to ease the satellite into its proper orbit, where it is now functioning normally.

The second and third launch failures occurred just hours apart on Shuttle mission 41-B in February 1984. The two satellites involved, Western Union's Westar 6 and Indonesia's Palapa B2, both employed McDonnell Douglas PAM-D upper stages for transfer to higher orbit. In their case the Thiokol STAR-48 solid fuel motors in each booster failed and both satellites ended up in useless elliptical trajectories. The mission was a public relations disaster for NASA, which is anxious to portray the Shuttle as a reliable commercial launch vehicle. Fortunately the story has a happy ending and ten months later the MMU (manned manoeuvring unit), which was tested for the first time on the same flight, provided NASA with the means to recover the situation. The exciting rescue of Westar and Palapa is described below.

Not all satellites require such complex methods of delivery. Many, particularly scientific research spacecraft, simply need to be left in low Earth orbit; and for these the Shuttle is provided with a simpler means of deployment. Tucked along one side of the payload bay is a sophisticated space crane known as the robot manipulator arm. Built by the Canadians, it is 15 m (50 ft) long and electrically operated, with flexible joints and a rotating wrist. It can grapple the largest payloads that the Orbiter is capable of carrying, and it can manoeuvre them in space with great precision. It can even be used to carry astronauts around the hold, providing a convenient work platform at almost any location.

The robot arm is controlled by a crew member working inside the Shuttle near one of the flight deck windows that overlook the payload bay. This is a demanding task and for some reason it seems to have become a speciality of female astronauts! A close-up TV camera and a clamping device are provided at the tip of the arm, allowing the operator to use the system almost as an extenion of his or her own body. The satellite to be deployed is first lifted from its cradle and then released gently into space. A small burst of fire from the Orbiter's reaction control jets puts a safe distance between the two spacecraft and the new satellite can be left to continue in its orbit.

The Shuttle's robot arm is controlled by an astronaut from inside the Shuttle. The gold-covered camera visible at the arm's 'elbow' provides a close-up view of what is happening at the business end.

**Rescue in space**

As a means of launching satellites, the Shuttle faces some stiff competition. Unmanned expendable boosters, like the European Space Agency's successful Ariane series of rockets, can deliver equivalent payloads to geostationary orbit more economically and without the need for potentially unreliable upper stages like IUS and PAM-D. But there is one respect in which the Shuttle leaves its rivals standing – and that is in its unique ability to rendezvous with existing satellites and either repair them in orbit or, if necessary, bring them back to Earth. To date the Shuttle has demonstrated this capability on no fewer than four occasions, and these missions have been among the most exciting and spectacular flights of recent years. The highlight was undoubtedly the joint rescue of Westar 6 and Palapa B2 in November 1984.

The fact that the mission took place at all was itself something of an achievement. Following the loss of Palapa and Westar earlier in the year, complex four-way negotiations began between the satellites' manufacturer (Hughes Aircraft), their owners (Western Union and the Indonesian government), their insurers (primarily Lloyd's of London), and NASA. At first there was little enthusiasm for such an untried and apparently risky undertaking.

Then in April 1984, astronauts on board *Challenger* successfully repaired Solar Max, a damaged astronomical satellite that had been in orbit for more than four years. This rescue was a triumphant demonstration of the technology necessary to retrieve the two misplaced communication satellites and it persuaded the insurers to put up $5.5 million towards the cost of the new mission. Lloyd's had already paid the bulk of the $180 million insurance claim following the initial loss, and so this was the insurers' chance to recover and resell $70 million worth of hardware. NASA, for its part, was anxious to restore confidence in the Shuttle as a launch vehicle, and so agreed to subsidize the remaining mission expenses.

So on 11 November 1984, Space Shuttle *Discovery* blasted off at the start of mission 51-A. More was at stake than simply the recovery of Westar and Palapa: the value of the manned space programme was being put to the test under very public scrutiny. In the days before launch, ground controllers had commanded the two orbiting satellites to fire their thrusters and drop down to an accessible altitude. Within 24 hours *Discovery* achieved rendezvous with Palapa, and astronauts Joe Allen and Dale Gardner climbed into their space suits in preparation for the rescue.

At this point two important pieces of space hardware enter the story: the MMU or manned manoeuvring unit and the Stinger. The MMU is a remarkable astronaut backpack that allows a Shuttle crew member to fly around untethered in the vicinity of the Orbiter. It is powered by high-pressure nitrogen gas issuing from 24 separate jets clustered at the four corners of the unit. The jets are operated by two hand controllers:

the right hand controls rotation (pitch, roll, yaw), and the left hand controls translation (motion forwards/backwards, up/down, left/right). For safety reasons all the main MMU systems are duplicated. There are 2 nitrogen tanks, 2 gas distribution lines, 2 sets of electronics, and the 24 thrusters are divided into two functional sets of 12. Even in the worst case – a thruster valve sticking open – the MMU could never build up enough velocity to outrun the Orbiter.

The MMU allows an astronaut to approach a spinning satellite but another device is required to make the capture itself. Most modern satellites are now equipped with a small docking peg for just this purpose, but unfortunately this was not the case with either Westar or Palapa, since neither were designed for service in orbit. So NASA's engineers had to rack their brains and come up with another idea. The result was a cunning device called the Stinger, which would allow an astronaut riding his MMU to capture a satellite like a knight on his charger lancing a dragon.

The rescue began as Joe Allen fired his MMU thrusters and jetted smoothly across to Palapa which was spinning slowly in bright sunlight. Aligning himself with the blackened nozzle of the satellite's failed rocket motor, Allen plunged the tip of his Stinger into the dark hole and operated the trigger. Like an umbrella opening inside a chimney, the expanding prongs of the Stinger gripped Palapa's insides and the astronaut found himself locked to the satellite in a surreal weightless spin. At the touch of a switch marked 'attitude hold', jets of nitrogen gas issued briefly from the MMU and all rotation ceased.

Allen's next move was to bring the satellite slowly round towards the waiting Shuttle where Anna Fisher (the first mother in space) was poised at the controls of the robot arm. Fisher then used the powerful arm to grab the Stinger and grapple Palapa over to where Dale Gardner was waiting patiently in the payload bay. At this point a problem arose. According to the carefully rehearsed plan, Gardner should now attach a special metal frame to the satellite that would allow it to be clamped securely into a cradle for the journey home to Earth. But a small metal panel on the satellite's antenna protruded about 2 mm beyond specifications and the frame simply would not fit. So instead Joe Allen shouldered the load. For 90 minutes, an entire orbit of the Earth, he held Palapa steady while Gardner struggled to secure it in position by hand. On the ground the satellite would weigh ten times more than the astronaut, but in space he could lift it with ease. Weightlessness does not eliminate inertia, however, and Allen was complaining of painful cramps by the time Gardner had pinned it securely in place.

Inspired by their success in recovering Palapa, the crew of *Discovery* prepared to repeat the performance with Westar. After 24 hours' rest, during which time the Orbiter manoeuvred to rendezvous with the second satellite, Allen and Gardner climbed back into their space suits and began another six-hour sortie. This time Gardner donned the MMU

Mighty Joe Allen keeps a firm hold on Palapa while Dale Gardner prepares to secure the captured satellite in the Shuttle's payload bay.

and flew over to spear Westar with his Stinger, while Allen moved over to the robot arm and prepared to try out a new method of recovery.

Once again Anna Fisher was at the controls back on board *Discovery*. This time she used 'mighty Joe Allen' as a human grappling hook: his feet firmly braced in supports at the end of the robot arm and his two outstretched arms grasping Westar. Using this technique, the second satellite was secured in its cradle an hour and a half ahead of schedule and the astronauts had plenty of time to look around and take in the view. In high spirits they unveiled a small sign above their captured treasure. It simply read 'For Sale'! (The two satellites were now the property of the underwriters, who later recovered part of their losses by reselling the refurbished spacecraft.) Down below at Lloyd's of London the famous Lutine bell was rung twice: the traditional signal of a successful maritime salvage operation.

**Counting the cost**

The spectacular rescue of the Westar and Palapa satellites was an impressive demonstration of what man can achieve in space and vivid proof of the Shuttle's unique capabilities. But from a strictly economic point of view it is questionable whether the exercise was worth while. Before recovering the two spacecraft, *Discovery* deployed two other satellites and this helped to offset the cost of the mission. But even so, the rescue was certainly not cheap. In the light of such a glowing

In high spirits after their successful rescue of Palapa and Westar, astronauts Gardner (left) and Allen advertise the salvaged satellites.

success it may seem mean to count the dollars; but the sad truth is that the Shuttle, for all its technological virtues, is an expensive way of operating in space.

It is almost impossible to disentangle the complex economics of the Shuttle and come up with a realistic estimate of what a single launch actually costs. Do you, for example, include the $10.1 billion (in 1984 dollars) spent since 1972 in developing the Shuttle; or the $9.3 billion spent in constructing the fleet of Orbiters and all their associated hardware? (Let alone the cost of replacing *Challenger*.) Even if you do decide to exclude these 'sunk' costs, it is now widely accepted that the $80 odd million that NASA charges customers for a Shuttle launch is considerably less than the real cost of a mission. Most independent estimates (for example in a recent study by the United States Congressional Office of Technology Assessment) put the figure at between $150 and $200 million per flight. For the purposes of launching satellites to geostationary orbit, the Shuttle is considerably more expensive than unmanned expendable boosters like Delta or Atlas-Centaur, which cost about $120 million for an equivalent payload.

Of course the Shuttle can also service and retrieve existing spacecraft, which is a facility not offered by any other system. But for the foreseeable future, the total number of repair and recovery missions is unlikely to exceed one or two a year. There are plans for the Shuttle to rendezvous with the Landsat 4 Earth resources satellite and replace a failed solar panel that currently renders that spacecraft virtually useless. But beyond this mission there are few economically worthwhile candidates for rescue within range of the Shuttle's low orbit.

The *Challenger* disaster has come at an unfortunate time for the Shuttle programme. Despite an impressive string of technological achievements, in economic terms the vehicle has largely failed to provide the low-cost, high-frequency launch service that was originally promised. Back in the early 1970s, when President Nixon approved its development, NASA was confidently estimating shuttle payload-to-orbit costs of $100 per pound and launch rates of 50 plus per year. In the event, the first figure has proved to be at least ten times too low and the second more than four times too high. Now with the loss of seven human lives and two billion dollars worth of hardware, these criticisms have become impossible to ignore.

1986 was to have been the Shuttle's busiest year with 15 important flights scheduled. This carefully planned timetable of missions has been thrown into disarray by the temporary grounding of the three Orbiters remaining after the loss of *Challenger*. The Shuttle is crucial to almost every scientific, military and commercial space project planned for the next decade and the short term outlook for all forthcoming missions is uncertain.

Important scientific missions that are scheduled to fly in the near future include Spacelab (a large orbiting scientific laboratory built by the European Space Agency, which is described in detail in Chapter Three) and several major scientific spacecraft including Space Telescope, the Galileo probe to Jupiter and the Ulysses mission to the Sun. (Galileo and Space Telescope are described in Chapters Four and Five.)

In the field of industrial applications, the Shuttle has a busy programme of experiments flying under special agreements that NASA has made with commercial organizations. In some cases, such as the McDonnell Douglas/Johnson & Johnson project to manufacture high-value pharmaceutical products in weightlessness, actual production takes place on board the Orbiter. Initially, most industrial experiments are riding on the Shuttle for free. But NASA hopes that in years to come several processes will have matured to the point where their developers are prepared to pay a high price for the ride into orbit. The growing commercialization of space is the subject of Chapter Six.

The Pentagon has a significant interest in Shuttle operations with almost 25 per cent of future flights booked by the Department of Defense. There is a new military Shuttle launch and landing facility at Vandenberg Air Force Base in California and a new cadre of Air Force astronauts have been trained to launch and service a wide range of military payloads – from early-warning spacecraft to high-resolution spy satellites. The Shuttle will also be vital for many experiments in President Reagan's Strategic Defense Initiative. Military activity in space is discussed in Chapter Seven.

Utah Senator Jake Garn: the first politician to hitch a lift on the Shuttle. During his flight doctors recorded the Senator's bowel sounds to discover whether they were affected by weightlessness.

NASA's policy of giving flight places to non-astronauts met with disaster when Christa McAuliffe, a 37 year old teacher and the first 'ordinary citizen' to ride on the Shuttle, was killed in the explosion that destroyed *Challenger*.

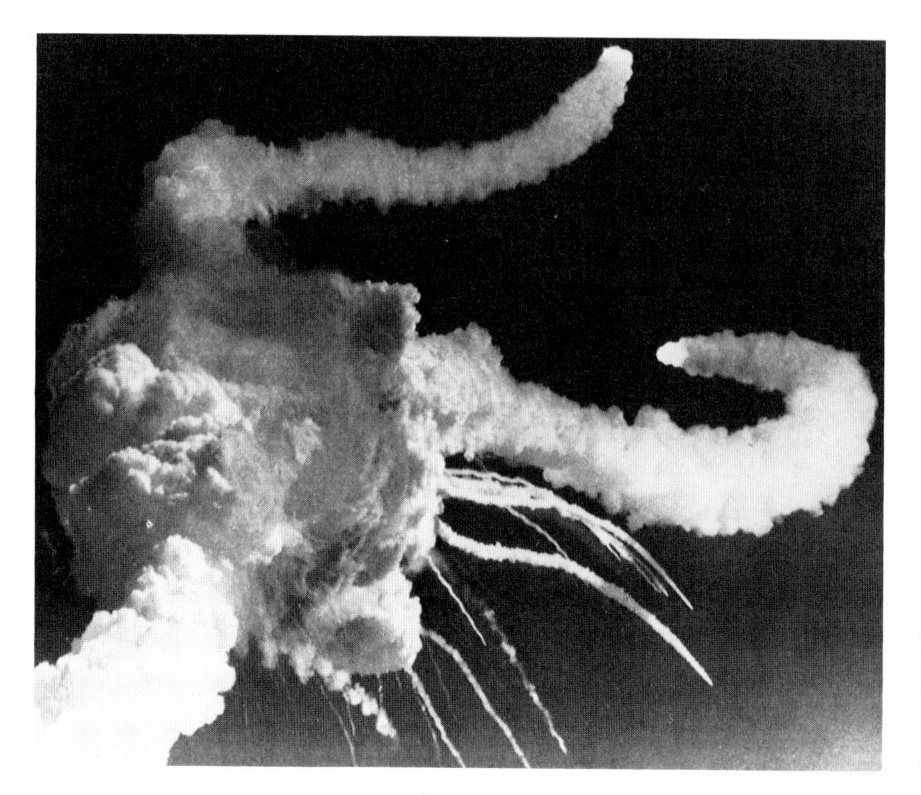

When *Challenger* exploded on 28 January 1986 the Shuttle programme was beginning its busiest ever year of operation.

This timetable of 20 Shuttle missions current at the time of the *Challenger* disaster indicates how seriously the programme will be disrupted.

**Launch manifest 1986-87**

| *Launch (D/M/Y)* | *Mission (code*)* | *Orbiter (name)* | *Duration (days)* | *Crew (size* | *Payload (primary)* |
|---|---|---|---|---|---|
| 15/05/86 | 61-F | Challenger | 2 | 4 | ULYSSES European solar polar mission |
| 21/05/86 | 61-G | Atlantis | 2 | 4 | GALILEO Jupiter probe/orbiter |
| 23/06/86 | 61-H | Columbia | 7 | 7 | EOS-2 commercial drug manufacture<br>PALAPA B-3 communications satellite<br>STC DBS-A communications satellite<br>SKYNET-4A UK military comsat |
| 15/07/86 | 61-I | Challenger | 7 | 7 | MSL-4 materials science laboratory<br>INSAT 1-C communications satellite<br>INTELSAT VI-1 communications satellite |
| 08/08/86 | 61-J | Atlantis | 3 | 5 | HUBBLE ST space telescope |
| 03/09/86 | 61-K | Columbia | 7 | 7 | SPACELAB/EOM-1/2 environmental observatory |
| 15/09/86 | 61-L | Challenger | 7 | 7 | DOD PAM-1 Department of Defense satellite<br>ASC-2 communications satellite |
| 29/09/86 | 62-B | Discovery | secret | secret | Second DoD mission out of Vandenberg AFB |
| 22/10/86 | 71-A | Atlantis | 7 | 5 | SHEAL-1 Shuttle high-energy astrophysics lab<br>SPARTAN-2 solar physics experiment<br>TDRS tracking and data relay satellite |
| 30/10/86 | 71-B | Columbia | 7 | 7 | MSL-5 materials science laboratory<br>SPACELAB/ASTRO-2 UV astronomy telescope |
| 05/12/86 | 71-C | Challenger | secret | secret | Department of Defense mission |
| 15/12/86 | 71-D | Atlantis | 7 | 6 | SPARTAN-3 ultraviolet imaging astronomy<br>DOD PAM-2 Department of Defense satellite<br>STC DBS-2 communications satellite<br>SKYNET-4B UK military comsat |
| 07/01/87 | 71-E | Columbia | 7 | 5 | MSL-6 materials science laboratory<br>INTELSAT VI-2 communications satellite<br>PAYLOAD OPPORTUNITY |

| | | | | | |
|---|---|---|---|---|---|
| 03/02/87 | 71-F | Challenger | 7 | 5 | LDEF-2 deploy long duration exposure facility |
| 25/02/87 | 71-G | Atlantis | 7 | 7 | SPACELAB/SLS-2 space life sciences |
| 09/03/87 | 71-H | Columbia | 7 | 5 | MSL-7 materials science laboratory<br>DOD PAM-3 Department of Defense satellite<br>PAYLOAD OPPORTUNITY |
| 02/04/87 | 71-I | Challenger | 7 | 5 | DOD PAM-4 Department of Defense satellite<br>DOD PAM-5 Department of Defense satellite<br>STACOM KU-4 communications satellite |
| 09/04/87 | 71-J | Atlantis | 7 | 5 | MSL-8 materials science laboratory<br>PAYLOAD OPPORTUNITY OR<br>INTELSAT VI-3 communications satellite |
| 07/05/87 | 71-K | Columbia | 7 | 7 | SPACELAB/IML-1 International microgravity lab |
| 09/06/87 | 71-L | Challenger | 7 | 5 | OAST-3 large solar array experiment<br>DOD PAM-6 Department of Defense satellite<br>CRRES atmospheric research satellite |

The Shuttle has another role which is common to all its missions and that makes the loss of *Challenger* such a serious blow to NASA. The Shuttle has inherited from earlier space projects, like Mercury, Gemini and Apollo, the political function of demonstrating the technological superiority of the United States and in that respect the Shuttle is a powerful symbol of national pride. This strange fact, so divorced from the practical realities of spaceflight, is partly why President Reagan has been such a strong supporter of NASA. It is the key to understanding the next major development in the American space programme which is the subject of Chapter Two.

For America the Shuttle is a symbol as well as a spacecraft.

CHAPTER TWO

# Space Station

*America has always been greatest when we dared to be great. We can reach for greatness again. We can follow our dreams to distant stars, living and working in space for peaceful, economic, and scientific gain. Tonight, I am directing NASA to develop a permanently manned space station and to do it within a decade.*

Ronald Reagan, 25 January 1984

President Reagan chose the occasion of his State of the Union address to declare the United States' next major goal in space. In a speech before Congress that was clearly intended to echo John Kennedy's 1961 commitment to go to the Moon 'before the decade is out', Ronald Reagan set the course of NASA's manned space programme for the rest of this century. The Space Station will cost at least $8 billion to build and many more billions of dollars to operate. It will become the focus of all American manned space activity as soon as it is assembled in orbit in the early 1990s. In fact 1992 will be the 500th anniversary of Columbus's landing in the New World and NASA would love to be ready in time for the celebrations.

### Designing a station in the sky

The main difference between the Space Station and all the other great projects that NASA has so far undertaken is that the Station has no

Two milestones in the history of the American space programme: John Kennedy's 1961 commitment to land a man on the Moon, and Ronald Reagan's 1984 commitment to build a permanently manned space station.

single clear overriding objective. Apollo existed to land men on the Moon and bring them safely home. The Shuttle was developed to provide a reliable and cost-effective transportation system to and from Earth orbit. But the Space Station is happening for a hundred different reasons. As a result NASA is taking an entirely new approach to its design, construction and eventual operation. NASA's philosophy in developing the Space Station is to regard the new project not as a piece of hardware with a particular fixed design, but as means of achieving a great number of different objectives.

An immediate consequence of this approach is that NASA is spending much longer than usual in the design phase of the Station's development. The intention is to establish very clearly all the things that the Space Station will be required to do before deciding on a precise configuration. To help guide the industry contractors who are due to begin constructing hardware in 1987, NASA has issued general guidelines for the overall Station architecture. The reference design is known as the 'Power Tower'. It gets its name from a long 140 m (450 ft) beam that will form the backbone of the final structure. The station will orbit the Earth with this main beam perpendicular to the ground. The top end will support large solar panels and radiators, while the bottom end will carry a cluster of pressurized modules where the crew will live and work.

Although many details remain to be worked out, one thing is already clear: NASA's facility will be utterly different from the popular image of a space station. For a start it will not be spinning. Probably the most famous space station design is the vast rotating wheel in Stanley Kubrik's film *2001: A Space Odyssey*. This magnificent conception has its origins in an article by the visionary German rocket designer Wernher von Braun called 'Crossing the Last Frontier' that appeared in *Colliers* magazine back in 1952.

At the time it was felt people would be unable to function in weightlessness, so von Braun set his ring-shaped station spinning in order to

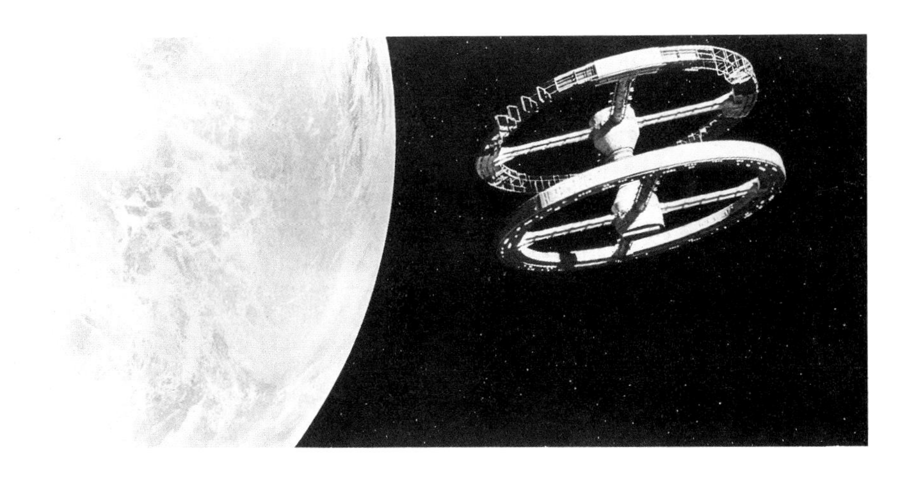

What NASA's space station will *not* be like: Stanley Kubrick's vision of a vast spinning cartwheel in his film *2001: A Space Odyssey*. *From the MGM release '2001: A Space Odyssey'* © 1968 Metro-Goldwyn-Mayer Inc.

One possible configuration for NASA's proposed space station: a visiting Shuttle is docked next to the cylindrical pressurized living quarters to the left. For a sense of scale see the small figure of an astronaut flying down towards the centre of the Station. An unmanned free-flyer platform is depicted at the lower right.

create artificial gravity for the inhabitants inside. Since those days, however, results from the long-duration missions on Skylab and Russia's Salyut have shown that man can adapt perfectly well to zero g. Indeed lack of gravity is generally a positive advantage because, among other things, it increases the effective size of a spacecraft interior by turning every surface into a potential floor.

Certainly the astronauts on board NASA's Space Station will need every cubic centimetre of room they can find, because the other important difference between science fiction and reality will be scale. At least during the first years of operation, the Space Station will be a relatively modest structure, weighing about 90,000 kg (200,000 lb), and with a crew of only six people. The Station will be constructed from prefabricated components carried into a low 500 km (310 miles) equatorial orbit by the Shuttle. At least seven Shuttle flights will be required to assemble the initial set of elements, but a key feature of the design is that there will be scope for considerable expansion in the future. In time the Space Station may grow to resemble our expectations more closely!

The most prominent feature of the Space Station will be a large array of solar panels to supply electricity to all the on-board systems. NASA is currently planning to generate as much as 120 kilowatts, which is enough to supply a small community on Earth and enormous by existing spacecraft standards. (The Shuttle, for example, has only 7 kilowatts of electrical power available.) Even using new high-efficiency gallium arsenide cells, the solar panels will extend to cover an area of more than 2000 $m^2$ (2390 $yd^2$), which is getting on for the size of half a soccer pitch. This large surface presents all kinds of control problems to

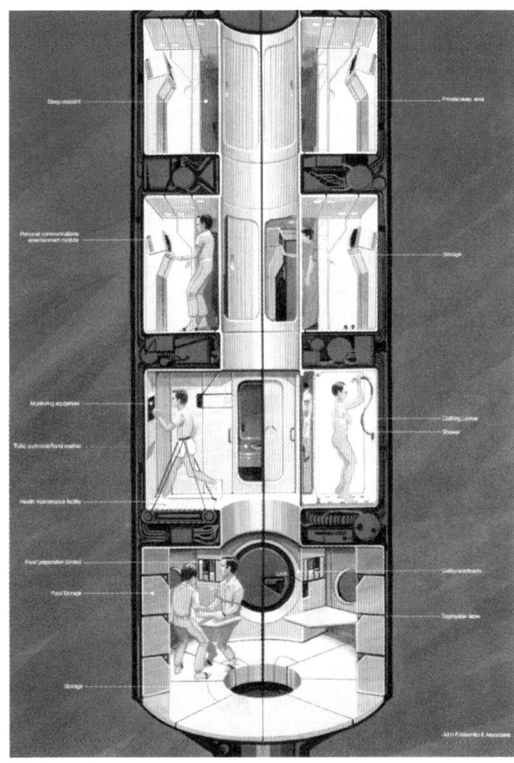


LEFT Large solar arrays, like this experimental unit tested recently on the Shuttle, will be used on the Space Station to supply large amounts of electrical power. The dark vertical line is a shadow cast by the Shuttle's tail fin.

RIGHT A possible configuration for the Space Station's 11 metre-long pressurized living modules. Life on board will be cramped.

the designers. As the Station orbits the Earth, the panels must slowly twist to keep facing the Sun. They must be strong enough not to snap and yet light enough to be carried into orbit.

The heart of the Space Station consists of several pressurized modules where the crew will live and work. These cylindrical modules will be limited in size by the capacity of the Shuttle's cargo bay, inside which they must fit for delivery from Earth. The bay is 18.3 m (60 ft) long and 4.6 m (15 ft) across, but economy will probably dictate shorter units about 10×4 m (33×13 ft). Each module will have at least one airlock connecting it to the rest of the Station and each will be specialized for a particular role. There will probably be two 'habitation modules', as NASA likes to call the crew sleeping, eating and resting quarters. In addition there will be a laboratory module for scientific experiments, an industrial module for manufacturing processes, a logistics module for life support and Station control, and probably other modules added as required.

Managing water and oxygen supplies will present new challenges to the Station's designers because in the long term they will have to develop a closed-loop system. All previous missions have carried enough supplies for the entire duration, discarding waste along the way; but with six crew members in permanent residence this option would be prohibitively expensive on the Space Station. A dedicated Shuttle mission would be required every six months just to resupply water. So instead, the Station's life-support systems will employ recycling technology to extract moisture from humid air and to distil pure drinking water from the crew's urine and bathing effluents. Electrical power could be used to regenerate oxygen from carbon dioxide, and there are even plans for the astronauts to grow fresh vegetables as Russian cosmonauts have successfully done on Salyut.

### A workshop in orbit

The third main component of the Space Station will be an open workshop and servicing area where all manner of exterior operations can take place. Docking facilities for the Shuttle will be close to this platform and the Station will have one or more robot manipulator arms to transfer cargo to and from the Shuttle's payload bay. Apart from supplies, the Shuttle will deliver new Station elements, satellites, scientific instruments, and raw materials for industrial processes. Out on the platform, space-suited astronauts will be able to work servicing and repairing satellites, assembling experiments and constructing new sections of the Station. Then, using Manned Maneouvring Units (MMUs), they will be able to fly the length and breadth of the entire structure, bolting on extensions like kids playing with a Meccano set.

The remaining principal elements of the Space Station will have no physical connection with the main structure. NASA has plans to include a number of unmanned platforms in the system that would orbit the Earth either in formation with the manned section, or else in their own independent orbits. These free flyers will house various experiments and devices which, for one reason or another, need to be separated from the main Space Station. Some of them would be placed in specialized orbits tailored to a particular function: a sun-synchronous orbit for solar observations, or polar orbit for remote-sensing applications. The free flyers will be serviced by the Shuttle which would visit them from time to time as the need arises.

In the longer term, NASA has more ambitious plans for the Space Station. High on the list of priorities is a small space 'tug' that would be

**Timetable**

| *Year* | *Mass (kg)* | *Volume (m³)* | *Power (kW)* | *Modules (*)* | *EVA (hours)* |
|---|---|---|---|---|---|
| 1991 | 35,900 | 195 | 55 | 4 | 830 |
| 1992 | 49,000 | 195 | 60 | 4 | 1450 |
| 1993 | 55,000 | 215 | 63 | 5 | 1520 |
| 1994 | 61,000 | 275 | 88 | 8 | 1330 |
| 1995 | 101,000 | 275 | 101 | 11 | 230 |
| 1996 | 107,000 | 310 | 106 | 14 | 160 |
| 1997 | 101,000 | 305 | 121 | 14 | 150 |
| 1998 | 99,000 | 305 | 112 | 14 | 140 |
| 1999 | 96,000 | 370 | 130 | 15 | 110 |
| 2000 | 94,000 | 365 | 129 | 15 | 70 |

One possible timetable of Space Station growth during the 1990s. NASA is exploring many different options and the above figures indicate only one of many alternatives.

garaged at the Station and could be used to collect and deliver satellites from orbits that are beyond the reach of the Shuttle. This orbital manoeuvring vehicle (OMV) would be unmanned. It would be flown using remote control by astronauts on board the Station, or even back on Earth, and it would have a robot arm that could be used to grapple defective satellites. The OMV would resemble a pancake, measuring roughly 3 m (10 ft) across and 1 m (3.3 ft) thick. It would be completely re-usable and it could reach orbits as high as 2000 km (1245 miles), which is several times higher than the Shuttle's ceiling. This new vehicle would significantly enhance the servicing and repair capabilities of the Shuttle/Space Station system.

Further into the future, NASA would like to develop a manned ferry vehicle that could travel even higher. This spacecraft, known as the orbital transfer vehicle (OTV), would be assembled on the Space Station and, since it would never have to fly in the atmosphere, it could be very light and highly efficient. Like the OMV, the OTV would be refuelled at the Space Station and it would have sufficient range to take astronauts right up to geosynchronous orbit. Here they could service and repair the vast fleet of satellites that occupy this vitally important region of space 36,000 km (22,000 miles) above the planet.

**Working in space**

The significant thing about the Space Station is not the hardware from which it will be constructed, but the uses to which it will be put. During 1982 and 1983, NASA commissioned eight major investigations by the leading aerospace companies: Boeing, General Dynamics, Grumman, Lockheed, McDonnell Douglas, Martin Marietta, Rockwell and TRW. The contractors were asked to find out if there was a market for the services of a space station and what tasks it might carry out. As a result of this study more than 100 different missions were identified. These Space Station missions fall into three principal categories: scientific research, technology development, and commercial operations.

The fact is that almost every space mission during the 1990s could potentially involve the Space Station in one way or another. In some cases the involvement would be considerable, with hardware and people actually present and working on board. In other cases the Station would simply be used as a staging post or assembly facility for independent spacecraft carried up on the Shuttle. At the very least, the Space Station would have an insurance role: ailing satellites could be brought in for service or repair in the event of failure.

In the science category, NASA has identified six general types of mission that would involve the Space Station: astronomy and astrophysics, Earth sciences, solar system exploration, life sciences, materials science, and communications research. Each of these areas of investigation will place different demands on the Space Station and each will have its own impact on the final design.

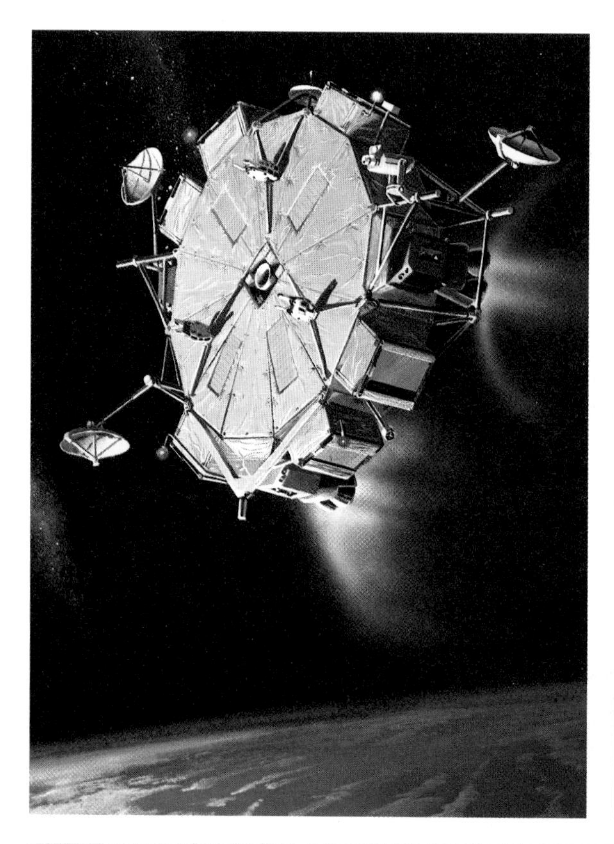

TOP LEFT An artist's conception of the OMV or orbital manoeuvring vehicle. This re-useable robot spacecraft will be employed to ferry satellites (attached by the three small clamps visible at the front of the craft) between low Earth orbits in the vicinity of the Shuttle.

TOP RIGHT Larger and more powerful re-useable vehicles will be required to reach geostationary and lunar orbits.

BOTTOM The Shuttle will be used to deliver elements of the Space Station in stages. The size of individual modules will be constrained by the capacity of the Shuttle's payload bay.

On-orbit satellite repair, first demonstrated with the successful refurbishment of the astronomical research spacecraft Solar Max, will be one of the activities carried out on board the Space Station.

Astronomical instruments and remote-sensing cameras for observing the Earth in great detail are likely to be located on unmanned free-flying Space Station platforms. These experiments demand very high pointing accuracy and they are sensitive to contamination. Isolation on free flyers will ensure that they are not disturbed by astronauts' movements or by the environmental pollution that life-support systems in the manned part of the Station will inevitably introduce. Some missions will have their own dedicated spacecraft in independent orbits – such as the Spacc Telescope, the Gamma Ray Observatory and the Advanced X-Ray Astrophysics Facility (*see* Chapters Three and Five). In these cases astronauts from the Space Station will periodically visit the instruments to maintain them or to exchange cameras and other devices. In the event of major failure, the entire spacecraft could be brought to the Space Station for more extensive repairs.

Unmanned planetary spacecraft will use the Space Station as a springboard out to other parts of the solar system. They will be carried up on the Shuttle and then prepared for departure by astronauts working outside on the service platform. The booster rockets that will carry the probes away from Earth could be fuelled with cryogenic propellants stored on tanks attached to the Space Station. In some cases, such as the proposed Mars Sample Return mission, large spacecraft will be assembled from smaller subsections delivered on several Shuttle flights. Astronauts working outside the Space Station will also have an important role in communications research, which is likely to involve the construction of large experimental antenna arrays.

Life sciences and materials processing investigations will take place on board the Space Station in purpose-built laboratory modules. Here specialist scientists, with a minimum of astronaut training, will be able to work for extended periods in orbit. They will investigate the effects of long-duration weightlessness on humans, animals and plants; they will study the growth of crystals in zero g; and they will investigate numerous processes that show commercial promise as the basis for potential space industries. All these experiments will benefit from the presence of humans to evaluate results and to notice unusual phenomena, changing and adapting methods and procedures with much greater flexibility than automated systems. NASA hopes that the Space Station will become a sort of research institute in orbit, a science park where an entire new industry can be developed to exploit the unique characteristics of space.

The second major category of Space Station missions will be in the area of technology development involving experiments, devices and procedures intended to enhance and extend future operations in space. These missions will place heavy demands on the Station's astronauts who will have to undertake extensive extra vehicular activity (EVA) to attach experimental devices and to try out new methods of construction, satellite repair, and materials handling. The astronauts will assemble a large solar reflector to concentrate sunlight for industrial furnaces, they will practise the techniques required to service and maintain the OTV space ferry, and they will experiment with laser communications, new types of rocket engine, closed-cycle life-support systems, and the deployment of very large structures in space. These missions will advance the frontiers of space technology and greatly extend the Space Station's capabilities and usefulness.

**Profit in the sky**

The third category of activities on the Space Station is one that NASA has been selling for all its worth because it appeals to President Reagan and to the politicians who must fund the programme. NASA believes that the Station has considerable commercial potential and that there are large profits to be made in space. They claim that government investment in the Space Station is justified because it will stimulate a vast new industry – one that will ultimately return great wealth to the United States.

So far only one space industry has been an outstanding commercial success and that is the telecommunications business. There are now more than 120 communications satellites in orbit serving 150 different countries and generating an annual revenue of over $10 billion. According to the most widely accepted estimates, a further 300 to 400 communications satellites will be launched between now and the year 2000 in the civilian sector alone. This represents an enormous market in which, NASA believes, the Space Station could have a major role to play. Satellites would be launched from the Station in much the same way as they are now deployed from the Shuttle. When the OMV and OTV space ferries become available, the Station would also become involved in maintenance and repair with a large inventory of spare parts on hand for speedy service. It would even be feasible to store entire back-up satellites on the Station ready for launch at short notice to restore communications in the event of sudden failure.

NASA believes that the telecommunications business points the way to a rich commercial future and that many space industries are ripe for development on the Space Station. The list includes Earth resource applications that would employ sophisticated cameras, lasers and radars to study the planet. The resulting data would be sold to governments and companies interested in mapping agricultural land use, mineral resources, pollution, fish stocks, crop disease, urban growth and a dozen other kinds of information.

However, the main commercial opportunity that NASA claims for the Space Station is in the field of materials processing. Weightlessness offers industry the chance to manufacture entirely new substances that cannot be produced on Earth. And zero gravity techniques promise enormous increases in the efficiency of certain production processes, particularly in the pharmaceutical and microelectronics industries. The possibilities here are indeed exciting and likely to have an important influence on the future in space. They are the subject of Chapter Six.

There is a fourth area which may be significant for future operations on the Space Station but, unlike commercial opportunities, it is not one that NASA is fond of discussing. The Pentagon is extremely active in space with a wide range of communications, navigation, early warning, electronic surveillance, and spy satellites in orbit. Support from the United States Air Force was a major influence in the decision to develop the Shuttle and up to 25 per cent of future Shuttle missions will be for defence purposes. With such a close connection already established between NASA and the Pentagon it seems natural that the Space Station will have significant military applications.

But despite the extent of Department of Defense activity in space, and despite their close association with the Shuttle, it seems the military are not particularly interested in the Space Station. Indeed, top DoD officials have gone out of their way to claim it would have few military advantages and they have opposed its development on the grounds that it will take away funds from things they would rather see done.

There are a number of reasons why the Pentagon feels less attracted to the Space Station than might be expected. For one thing the Station will be extremely vulnerable in wartime. It will be in a widely known low orbit and unable either to manoeuvre or to defend itself against attack. In peacetime, the Station will be in the full glare of the world's media: there could hardly be a worse place to conduct secret military activities. Furthermore, the place is likely to be crawling with foreigners! NASA hopes that many Western nations will be contributing crew members and Station components.

Indeed, European plans to participate in the Space Station programme are well advanced. Following President Reagan's invitation, the European Space Agency (ESA) has agreed in principle to provide its own pressurized manned module, known as *Columbus*, that would be attached to the main structure. Unlike NASA's modules, *Columbus* would be largely self-sufficient: carrying its own life-support equipment, power supply and propulsion system. This feature would make it relatively easy for Europe to develop its own independent station. ESA has also agreed to supply a number of free-flying platforms, probably based on the *Eureka* design manufactured by British Aerospace.

**The orbital marketplace**

However, the main main reason why the US Defense Department is not keen on the Space Station is a general one that has implications for many civilian applications as well. The Station will be in what is technically known as a 28.5 degree inclination, low Earth orbit. It will fly at an altitude of about 500 km (310 miles), travelling west to east at a slight angle to the Equator. This orbit is the one that is easiest and least expensive to reach from the launch pad at Kennedy Space Center in Florida. It takes maximum advantage of the extra boost provided by the rotation of the Earth and it means that the Shuttle can carry the largest possible payload to the greatest possible height. Unfortunately this 'cheapie' orbit has few other attractions to recommend it.

Polar orbit and geosynchronous orbit are the ones that really matter and they are used by the majority of the world's satellites. In polar orbit a spacecraft flies north-south (and south-north) at a fairly low altitude, more or less parallel to lines of longitude. As the Earth turns below on its axis, the satellite will, in due course, cover every part of the globe. Polar orbit is ideal for surveying the ground and most Earth resources and military surveillance satellites use it. In geosynchronous orbit, on the other hand, a spacecraft flies west-east, parallel to and exactly above the Equator. Geosynchronous orbit is much higher, at an altitude of almost 36,000 km (22,000 miles), and at this distance a satellite takes precisely 24 hours to make one circuit. Since the Earth also revolves exactly once during this period, the satellite appears at a fixed point overhead relative to an observer on the planet. This characteristic is perfect for weather, communications and early warning satellites.

Now the Space Station, stuck at an altitude of only 500 km (310 miles) and flying roughly along the line of the Equator, is pretty much cut off from both polar and geosynchronous orbit. Reaching the latter requires the expenditure of almost as much energy again as was originally required to lift a payload up from the ground. Additional boosters must be used to complete the second stage of the journey and this introduces delay, as well as an extra opportunity for something to go wrong. (Indeed, as described in the previous chapter, the Shuttle has experienced failures during the deployment of several satellites using exactly this technique.) Matters will improve when the OTV vehicle becomes available to ferry payloads from the Space Station up to geosynchronous orbit. But even so, the military prefer the immediacy and the security of a one-shot-to-orbit expendable booster, completely bypassing any involvement with the Space Station.

The argument is just the same for polar orbit, which is particularly important to the military because it is here that spy satellites must be deployed at short notice and with high reliability. Polar orbit is completely inaccessible to the Space Station because the energy required to change velocity through 90 degrees (which is in effect what would be involved) is almost twice that required to launch a payload in the first place! Space Stations systems that require placing in polar orbit will be

NASA has used outrageous analogies with the Wild West, as in this bizarre montage, to justify its 'Space Station on the New Frontier'.

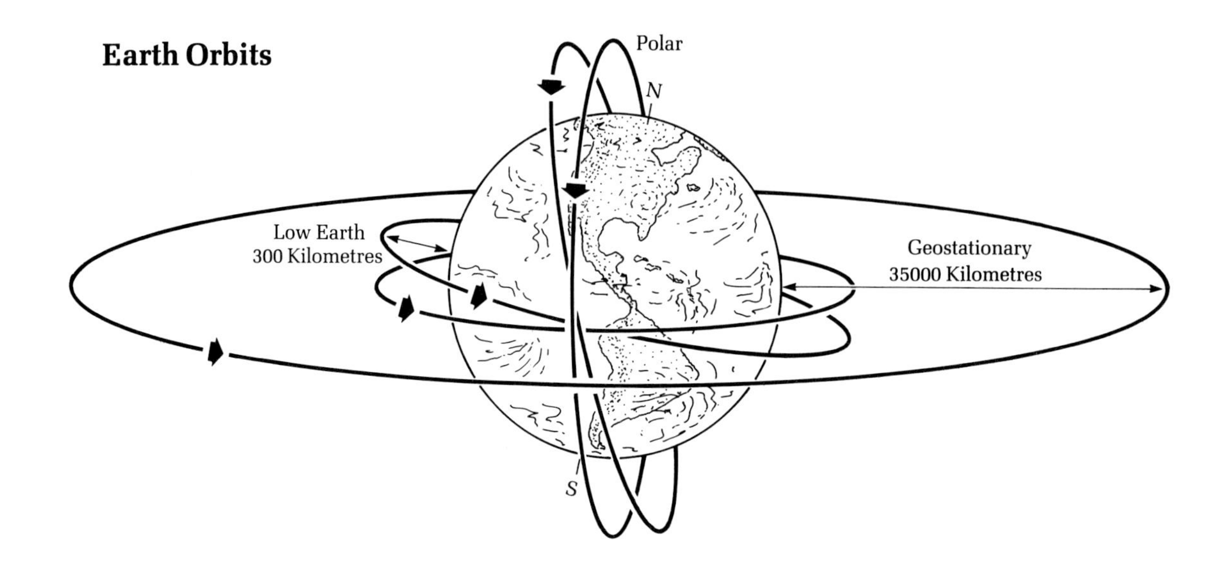


There are three economically significant Earth orbits: Low-Earth, Polar and Geostationary (sometimes called Geosynchronous). The Space Station will have direct access only to the first of these.

located on unmanned free-flying platforms, which will be delivered and serviced entirely by Shuttles launched from Vandenberg Air Force Base (*see* Chapter Seven). In effect, until NASA builds a polar-orbiting manned Station in the distant future, all polar operations will be completely independent, having nothing whatsoever to do with the astronauts on board the main Space Station.

Precisely these criticisms have also been voiced by prominent members of America's space science community who fear that the Space Station may dominate NASA's future operations to the detriment of scientific missions. They argue that unmanned spacecraft in a great variety of orbits would achieve the same scientific ends for much less cost. They claim that the vast majority of experiments and observations can be carried out more efficiently and more reliably with automated robot systems.

## The next logical step?

NASA has also been criticized for the hard sell it has given the Space Station's commercial prospects. There may well be some highly efficient industrial processes that depend on weightlessness, or unique and valuable substances that can only be manufactured in zero gravity. But the mere existence of such possibilities does not mean that it makes economic sense to build a Space Station in which to carry them out. As discussed more fully in Chapter Six, the cost of operating in space is very high and the commercial future for materials processing is still far from clear. In the current political climate, the claim that the Space Station will bring economic benefit back to Earth is certainly popular. Some analysts, however, believe that NASA may be making a big mistake in claiming benefits for the Space Station that it will simply be unable to deliver.

Unofficially, NASA is prepared to admit that there is a plenty of truth to these criticisms. The package of justifications it has developed to win political approval for the Space Station has been something of a 'kitchen sink' advertising campaign: every conceivable application has been thrown in without much hard-nosed evaluation but NASA believes that the Space Station is not, and should not be justified purely in terms of a few foreseeable practical benefits. NASA's leaders argue that the Space Station should be built because it is the way forward to the future: 'the next logical step in space' as they like to say at every opportunity. In this context, NASA officials tell the story of the mayor of a small American town who was shown a telephone for the first time back in the last century. 'This is a fantastic invention,' he said, 'I foresee the day when there will be one in every big city!' Well, NASA believes that the Space Station will be just like the telephone, making possible a rich future that it is impossible for us to predict today.

Back in the heady days of 1969, when Neil Armstrong was about to set foot on the Moon and when the space adventure was firing the imagination of the world, NASA's leaders were not afraid to express what must ultimately be the true justification for building the Space Station. Thomas Paine, acting administrator at the time, wrote to President Nixon urging him to approve development of an ambitious 50-man space station which NASA was proposing for the 1970s:

> There are many potential valuable uses of such a space station, and new ones will be found as experts in many fields become familiar with the possibilities and are able to visit and actually use it. However, we believe strongly that the justification for proceeding now with this major project as a national goal does not, and should not be made to depend on the specific contributions that can be foreseen today in particular scientific fields like astronomy or high energy physics, in particular economic applications, such as Earth resource surveys, or in specific defence needs. Rather, the justification for the space station is that it is clearly the next major evolutionary step in man's experimentation, conquest, and use of space. The development of man's capability to live and work economically and effectively in space for long periods of time is an essential prerequisite not only for operations in Earth orbit, but for long stay times on the Moon and, in the distant future, manned travel to the planets.

As it turned out, NASA's vision of the post-Apollo era was firmly rejected. The result has been a decade of decline in the manned space programme as the under-funded Space Shuttle struggled to get off the ground. But following that experience, there is once again a sense of optimism at NASA's space centres. The Space Station is the gateway to an exciting future and although NASA is currently playing down the long-term implications, there is no doubt that the project heralds a renaissance in space. Some of the more dramatic missions that the Space Station will make possible – a permanent base on the Moon, mining expeditions to the asteroids, a manned expedition to Mars – are the subject of the last chapter in this book.

NASA
NASA

CHAPTER THREE
# Science in Space

*As long as there have been humans we have searched for our place in the Cosmos . . . We find that we live on an insignificant planet of a humdrum star lost in a galaxy tucked away in some forgotten corner of a universe in which there are far more galaxies than people. We make our world significant by the courage of our questions and by the depth of our answers.*

Carl Sagan, astronomer

When a space scientist gazes into the heavens on a starry night he sees an enormous laboratory. Out in space answers can be found to some of the most sublime questions that can be asked, questions concerning the nature of space and time, the life and death of stars, the origin of planets and even the creation of the Universe itself. Space science is a key source of new ideas in physics, with a regular flow of important astronomical discoveries that challenge established wisdom and demand new explanations. Space science is also concerned with more practical affairs: the study of Earth's atmosphere and climate so that we can understand the impact of our activities on the environment; the investigation of human performance in weightlessness so that astronauts can work more efficiently and avoid becoming sick; and the perfection of new industrial techniques to make unique and valuable products in orbit.

Scientific research continues to be one of the principal reasons why man and his machines venture into space. Space science can be conveniently divided into three areas of operation: manned research requiring the participation of astronauts; planetary exploration using

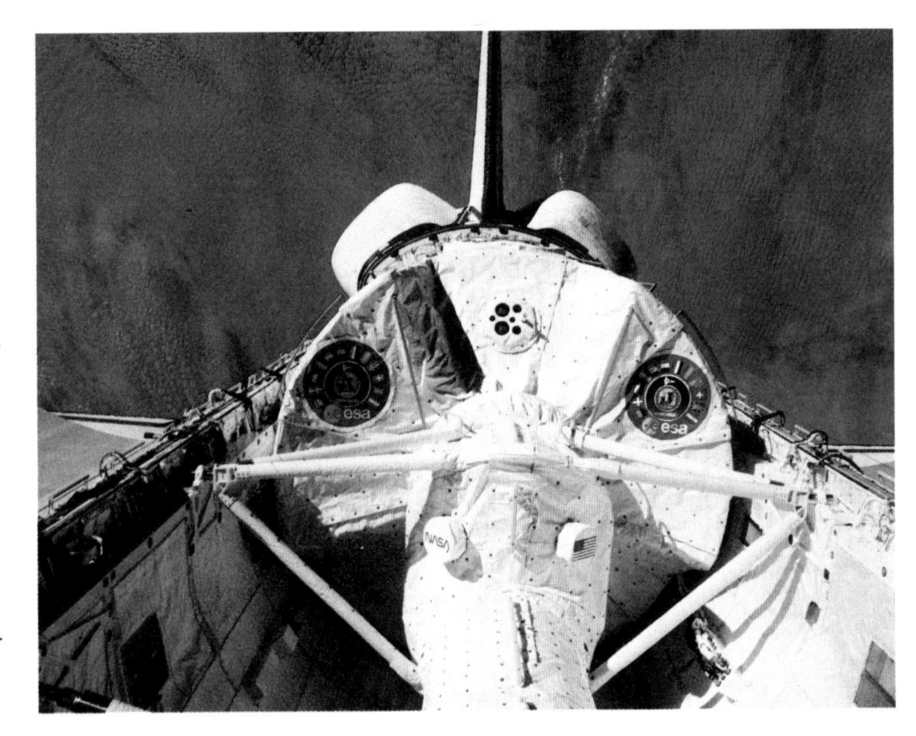


The crew of Spacelab 1: Their week in space was the fruit of lengthy and expensive collaboration between Europe and America – but was the scientific return worth all the effort?

Spacelab's main pressurized module is situated towards the rear of the Shuttle to maintain an acceptable centre of gravity. Access is via a pressurized tunnel from the Orbiter's crew compartment.

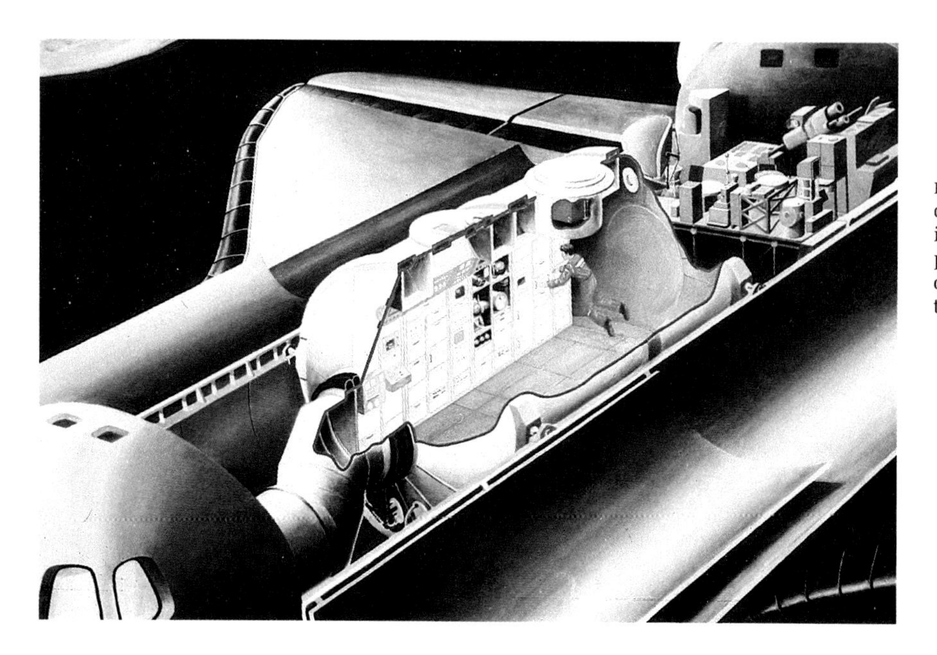

LEFT Spacelab experiments can be housed on racks inside the manned pressurized module, or outside on special pallets in the payload bay.

unmanned robot spacecraft; astronomical missions that employ orbiting satellite observatories. A great variety of projects are planned in all three categories during the next decade, including two – Galileo and Space Telescope – that are so important they deserve chapters all to themselves. Increasingly space science is becoming an international undertaking and many of the missions described in this chapter are joint ventures between two or more countries. This trend is partly happening for financial reasons, but scientific cooperation in space is also seen as a political tool for strengthening alliances back on Earth.

## A laboratory in the sky

Spacelab is the West's primary means of carrying out manned research in space and it will continue to dominate the picture until Space Station becomes fully operational in the mid-1990s. Spacelab is a system of manned and unmanned modules that fit in the Shuttle's payload bay, converting the Orbiter from a simple delivery vehicle into a sophisticated and highly adaptable laboratory in space. Most of the flight hardware for Spacelab was built by the European Space Agency, and ESA continues to collaborate with NASA in many Spacelab missons.

NASA has two complete Spacelab 'kits' that can be installed in a variety of configurations in any of the Shuttle Orbiters. The main component is a habitable pressurized module which provides a working environment for the Spacelab scientist–astronauts who are known as payload specialists. The cylindrical laboratory modules are 4 m (13 ft) in diameter and come in two different lengths: a single-segment version 4 m (13 ft) long, and a double segment version 7 m (23 ft) long. The larger version has a usable volume of 22 m$^3$ (777 ft$^3$) and it can carry up to 4600 kg (10,140 lb) of experiments. To preserve the Shuttle's centre of gravity, the pressurized module is located near the middle of the payload bay and the astronauts must reach it by floating down a narrow tunnel that connects to a hatch in the Shuttle's mid-deck area.

BOTTOM Various configurations of manned and unmanned Spacelab modules.

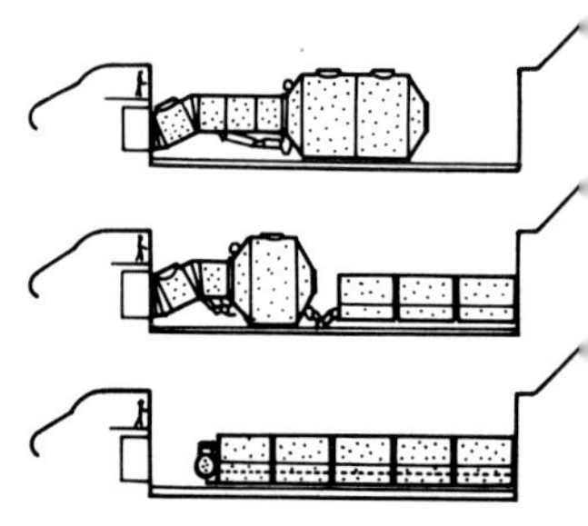

In addition to the pressurized modules, the Spacelab system includes a number of open pallets that fit neatly into the payload bay and which support experiments and instruments requiring direct exposure to space. These pallets are rigid U-shaped structures just under 3 m (10 ft) long and they each have numerous attachment points for securing up to 3000 kg (6600 lb) of scientific cargo. Pallets can be preloaded with experiments and then installed quickly into the Shuttle, minimizing Orbiter ground processing time. There is room for up to three pallets in the payload bay, although a more typical Spacelab configuration would consist of one long pressurized module and a single open pallet, or a short pressurized module and a pair of open pallets.

Two further devices complete the Spacelab system. The igloo is a telephone booth-sized service unit containing computers, electronics and other essential support systems normally carried in the pressurized modules. It is required on Spacelab missions employing only open pallets. The final large item of hardware is a sophisticated instrument pointing system. The IPS, which is housed on its own pallet, can be loaded with telescopes and other devices for observing both the heavens and the Earth. It can accommodate numerous heavy instruments and keep them accurately pointed at a target with a precision better than one arc second: quite an accomplishment given the notorious vibration and instability of the Shuttle in orbit.

Although Spacelab is now fully operational, it has had rather a troubled history. The European Space Agency developed the hardware at a cost of some $1000 million, in return for which ESA was permitted to select half the payload in the Spacelab 1 mission. After the mission was over, Spacelab became the property of NASA who also bought a second Spacelab system from ESA. A number of European scientists feel they got a distinctly raw deal, the more so since Spacelab 1 was essentially an engineering flight to test the system. Furthermore, development problems with the IPS, and more particularly with the Shuttle itself, have repeatedly delayed early missions. One unfortunate European scientist, who spent no less than seven years waiting for his experiment to fly, was prompted to come up with a new definition of lift-off: the moment when the weight of the payload equals the weight of the paperwork associated with it!

Despite these tribulations, however, Spacelab is now a versatile and effective means of carrying out manned experiments in almost any branch of space science. Inside the pressurized laboratory, scientists with a minimum of astronaut training can work in a shirt-sleeve environment and can gain first-hand experience of research in weightless conditions. The pressurized module is used as a place to set up and carry out experiments, as a home for plants and animals, and to store equipment and samples. There is a high-quality optical window for taking photographs of Earth and even a small airlock for exposing experiments to the vacuum of space. The laboratory also serves as a control room for instruments, such as telescopes and cameras, located outside on the open pallets. The only serious restriction is the necessarily brief duration of a Spacelab mission – limited by the present orbital endurance of the Shuttle to no more than 30 days.

The first Spacelab mission, which flew in November 1983, carried over 70 experiments spread across 6 different disciplines: materials processing, plasma physics, upper atmosphere research, astronomy, Earth resources, and the life sciences. One of the lessons that NASA learned from Spacelab 1 was that this sort of 'science cocktail' is a mistake. Conflicts arise, for example, over the required orientation of the Shuttle: payload bay facing the heavens for astronomy and facing the ground for Earth observations. Another problem concerns unrealistic demands placed on the payload specialists who fly on missions with such wide objectives. They must be jacks of all trades, trained in subjects as diverse as solar physics and plant metabolism.

Future Spacelab missions will specialize in particular branches of space science. Sunlab, for example, is currently slated to fly in late 1986. It will be a low-cost mission devoted to detailed solar observation and it is a good example of the flexibility that is possible with the Shuttle/Spacelab system. Sunlab will employ a single open pallet fitted with several telescopes and other instruments reflown from Spacelab 2. This configuration leaves plenty of room in the Shuttle's payload bay for two commercial satellites to be deployed shortly after launch. The crew will then be free to exploit several days of economical observation time – the overheads for the scientific part of the mission having been covered by the commercial satellite delivery charges.

### Sick as an astronaut

One of the top research priorities aboard Spacelab is space medicine. It may seem surprising, but a quarter of a century after Yuri Gagarin became the first human being to orbit the Earth, doctors still do not fully understand the effects of extended weightlessness on the health of astronauts. The medical problems that arise can be divided into short-term effects that become manifest just hours or even minutes after launch, and long-term effects that only become significant after several weeks or even months in zero gravity. Both types of problem are cause for concern. In the first case, astronauts' performance during the first crucial hours in orbit can be severely affected by space sickness. In the second case, astronauts may discover that their bodies are gradually weakened by prolonged exposure to weightlessness: even to the point where a return to normal gravity would be dangerous.

Space sickness, or space adaptation syndrome as it is euphemistically termed, is an unpleasant and debilitating condition that affects around half of all space travellers during their first few days in orbit. NASA, anxious not to tarnish the 'right stuff' image of its astronauts, has tended to play down the severity of the problem over the years. But it is

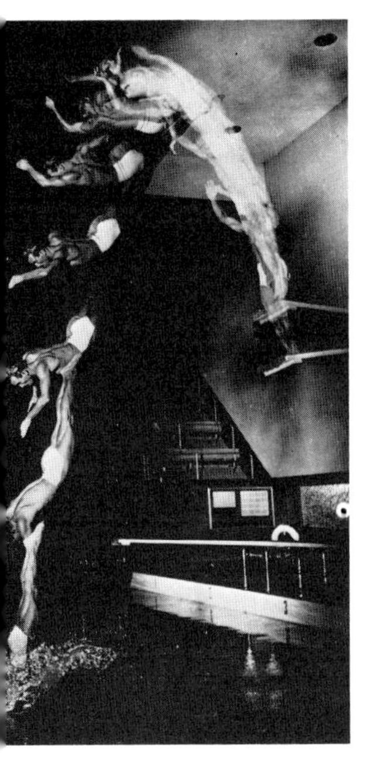

LEFT Contrary to popular belief, weightlessness is more akin to falling than floating (at least until your nervous system becomes adapted to the sensation).

RIGHT Spacelab 1 astronauts John Young and Ulf Merbold have conflicting definitions of up and down. When they meet, disorientation and nausea can result.

now clear that space sickness is not something that can be ignored or laughed away. The performance of almost every Shuttle crew has been affected with the result that precious time has been lost from busy work schedules. And space sickness is likely to become even more of a problem in the future as increasing numbers of people who are not highly trained career astronauts venture into orbit.

Space sickness is another form of motion sickness, similar from the victim's point of view to being seasick, carsick or planesick. At first, affected astronauts feel dizzy, disoriented and nauseous. If they are able to rest and lie still at this early stage then symptoms may subside; but if they are obliged to continue working and moving around (as astronauts often are during the first busy hours in orbit) then the malaise usually gets worse. Victims break out in a cold sweat, feel tired, lacking in initiative and extremely unwell, and eventually they vomit. One of NASA's greatest concerns is that space sickness could cause an astronaut to vomit while weightless inside a spacesuit. At the very least, the suit's life-support system would be badly damaged. At worst, the astronaut could inhale his own vomit and suffocate. Space sickness not only reduces a crew's efficiency: it also threatens life.

At first sight it seems surprising that 'floating' around in weightlessness could be anything other than delightful – but the sad truth is that it is initially extremely disturbing. About the closest thing to being weightless that we experience on Earth is during the few seconds between jumping off a high diving board and hitting the water. Being weightless feels exactly like falling, hence the term 'free-fall' that is sometimes used to describe conditions in an orbiting spacecraft. Since our brains normally experience the sensation of falling only for a few brief (and usually rather worrying) moments, it is perhaps not surprising that the illusion of continuous descent makes an astronaut feel unwell. However, the falling feeling itself usually wears off pretty soon after reaching orbit. It is the more subtle effects of weightlessness that lie at the root of the problem.

Thorough investigation of space sickness as it actually occurs in orbit has taken place on a number of the early Shuttle flights, including the first Spacelab mission in November 1983. Two of the Spacelab crewmen wore special devices to measure their head movements and they also dictated notes reporting how they felt into pocket cassette recorders. It soon became clear that two factors are particularly likely to make an astronaut feel sick: almost any type of head motion, and sudden spatial disorientation. During the course of the mission, three out of the six crew experienced persistent general discomfort, ate very little and vomited repeatedly. Despite these symptoms they carried on with their work – but only through sheer determination, exceptional motivation and true grit!

Confirmation that head movement tends to provoke nausea comes as no great surprise to space doctors. In weightlessness the sense organs in the inner ear that help us to balance and to detect motion are severely disturbed. This is because they rely on the presence of gravity for their normal operation. Familiar head movements, such as turning around to look at something, produce a flood of unfamiliar signals from our inner ear. According to this picture of space sickness, called the sensory conflict theory, our brain is confused by the unexpected messages and malaise is the result.

Paradoxically, the sensory conflict theory suggests that the best way for an astronaut to overcome space sickness is deliberately to experience the very head movements that make him feel unwell. Eventually the brain will learn to expect the abnormal signals and the feeling of nausea will subside. This is exactly what the Spacelab crew observed. After three days in orbit most of the crew were able to move around without feeling unwell – except when they were required to make unusual or particularly vigorous head movements. There is some evidence that anti-sickness drugs (which the astronauts can take to provide temporary and partial relief from nausea) may actually delay this process of adaptation. Crew members who resorted to medication took longer to overcome space sickness than equally affected colleagues who elected to vomit.

The second factor that tends to provoke space sickness is sudden spatial disorientation, as for example when two astronauts meet one another upside-down. During such an encounter there is apparently an overwhelming tendency to feel that you are the one who is the 'wrong' way round and about to fall, and this brings on an unpleasant rush of nausea. A similar phenomenon occurs when crew members float feet

Back to Earth with a bump: Cosmonauts Leonid Kizim (left), Vladimir Soloviov (centre) and Oleg Atkon (right) come to terms again with gravity after a record breaking 237 days in orbit.

first 'down' the tunnel linking the Shuttle flight deck to Spacelab's pressurized module. Everyone prefers to float 'up' the tunnel head first to avoid the powerful illusion of falling. On arrival in the relatively unconfined laboratory, astronauts agree to observe a common 'up' which follows their experience training in Spacelab simulators on the ground.

**Brittle bones and a wasted heart**

Space sickness is the most noticeable short-term medical consequence of weightlessness but there are more insidious effects that only become apparent after long spells away from Earth's gravity. American experience of these problems has come mainly from the Skylab missions back in the early 1970s when Gerry Carr, Ed Gibson and Bill Pogue captured the US endurance record with 84 days in orbit. Soviet cosmonauts have stayed in their Salyut space station far longer: Leonid Kizim, Vladimir Solovyev and Oleg Atkov hold the world (or should it be space?) record after a remarkable flight lasting 237 days in 1984. But after such prolonged exposure to weightlessness the cosmonauts were virtually stretcher cases for the first few days after their return to Earth.

The main problem on really long missions is lack of exercise. Not only is there little opportunity for vigorous activity inside a spacecraft, but the human body has a much easier time of things anyway in weightlessness. It has sometimes been said that a legless person would actually be at an advantage in space because astronauts make so little use of their lower limbs. The heart also has much less work to do pumping blood in the absence of gravity, and as a result its muscular walls gradually become thinner. This loss is restored after several weeks back on the ground, but there is concern that astronauts might experience heart failure on return to Earth after really prolonged missions.

French astronauts Jean-Loup Chrétien (on board Salyut) and Patrick Baudry (on the Shuttle) have used ultrasound to study the behaviour of the cardiovascular system in weightlessness. In the absence of gravity a major shift of fluid occurs from the arms and legs into the head and torso. This actually gives astronauts' faces a rounded, bloated appearance in orbit and it may well cause a number of less obvious and more sinister metabolic changes. The French device is able to produce images of the heart and arteries as blood flows through the body, rather like a moving X-ray. Studying changes in the distribution of body fluids during the first few days in orbit may help doctors to devise ways of reducing the extent of fluid shift in weightlessness.

Another serious medical consequence of long-duration spaceflight is loss of calcium from the skeleton. This effect can be measured even in Shuttle astronauts who have been away from Earth for only a week, and the process seems to continue inexorably with longer missions. Doctors believe that in the absence of gravity, the skeleton receives few bumps and shocks and almost no compression. As a result, astronauts returning home from several months in orbit find themselves a few centimetres taller, but with weak and brittle bones as if their skeleton had aged several decades. In fact some of the Skylab crew showed bone loss equivalent to natural deterioration in a person of 80 or 90!

Fortunately, most of the bone loss is restored after a period of recovery back on the ground, but again there is concern that a point may be reached on really long missions when it would be unsafe for an astronaut ever to come home. His skeleton would no longer be able to support him and he could suffer a broken leg simply from trying to walk down the stairs of the Shuttle. These problems do not affect current missions but they assume considerable importance for planning Space Station. The Soviet strategy on the marathon Salyut missions has been to enforce a rigorous and time-consuming programme of exercise lasting up to three hours per day. But even then, as photographs of the returning cosmonauts clearly reveal, the deleterious effects of prolonged weightlessness are only partially controlled. As one Russian remarked – back on Earth even a feather bed feels uncomfortable after months without gravity.

**Man versus machine**

Spacelab's great virtue is that it permits scientists themselves to venture into space and carry out experiments. Manned missions, however, are only part of the whole picture: unmanned spacecraft continue to be the mainstay of space science. Indeed, some space scientists would be happy to dispense with astronauts altogether! One of NASA's longest-running internal battles concerns the relative priority that should be assigned to manned and unmanned missions. For years a significant section of the science community have been sceptical about the value of man in space. They argue that experiments and observations are more efficiently carried out using automated equipment and robot vehicles that can be controlled from the ground.

In reply to such heretical suggestions, the supporters of manned spaceflight point out the ability of astronauts in orbit to respond to unexpected situations and to intervene when things go wrong. There

What is the best way of doing science in space: a cheap robot or an expensive monkey in a suit?

were numerous problems with equipment on board the first Spacelab mission, for example, that were corrected by the timely intervention of the crew. At one point, an important large-format camera for Earth resources photography jammed. It was repaired by an astronaut who was able to talk directly to the German technicians on the ground who had built the equipment. Similarly, during the spectacular Shuttle mission 51-A, it was the ability of astronauts Joe Allen and Dale Gardner to change plans in the face of circumstance that led to the successful rescue of the Westar and Palapa satellites.

But astronauts do not come cheap! Manned missions are expensive, not only because of the extra weight of the crew and their life-support system, but also because manned spacecraft must be built to the highest possible safety standards. Costly manned projects, such as the Shuttle and Space Station, consume the lion's share of available resources and starve far less expensive, though far more worthy, unmanned science missions. Yet without the excitement of manned operations to capture the taxpayer's imagination, it is doubtful whether space activity as a whole could be funded at the level it now is. Space science currently exists in an uneasy alliance with manned space operations: the former helping to justify the latter, and the latter allowing space science to proceed at a constant though fairly modest level for the rest of the decade.

### Last stop on the Grand Tour

Although the future is less rosy, 1986, at least, promises to be a truly vintage year for space science. The first highlight should be Voyager 2's encounter with the planet Uranus, 3 billion km (1.9 billion miles) away,

**Forthcoming attractions**

| *Year* | *Mission* | *Agency* | *Purpose* | *Description* |
|---|---|---|---|---|
| 1986 | Spacelab | NASA/ESA | Manned | Research in Earth orbit |
| | Voyager 2 | NASA | Exploration | Uranus encounter |
| | Giotto | ESA | Exploration | Halley's comet |
| | Vega 1 & 2 | USSR | Exploration | Halley's comet |
| | MS-T5/Planet A | Japan | Exploration | Halley's comet |
| | Ulysses | ESA | Exploration | Solar polar flyby (formerly ISPM) |
| | Galileo | NASA | Exploration | Jupiter probe and orbiter |
| | Space Telescope | NASA/ESA | Astronomy | 240 cm optical telescope |
| | RACAS | USSR | Astronomy | 30 m orbiting radio dish |
| 1987 | LBI | USSR | Astronomy | 10 m follow-on to RACAS long baseline interferometer |
| | CSMT | USSR/France | Astronomy | 1 m 49 K+1K detector microwave spectrometer/telescope |
| | Rosat | FDR/UK/NASA | Astronomy | 80 cm X-ray/XUV telescope |
| | Astro-C | Japan/UK | Astronomy | X-ray spectra |
| 1988 | Hipparcos | ESA | Astronomy | 30 cm optical telescope for determining star positions |
| | GRO | NASA | Astronomy | Gamma ray observatory |
| | Cosmic background explorer | NASA | Astronomy | Structure and spectrum of cosmic mircrowave background |
| | Venus radar mapper | NASA | Exploration | Venus surface imaging |
| | Mars probe | USSR | Exploration | Mission to land on Martian moons |
| 1989 | Voyager 2 | NASA | Exploration | Neptune encounter |
| | Lunar polar orbit | USSR | Exploration | Lunar geoscience mapper |
| 1990 | Mars Orbiter | NASA | Exploration | Mars geoscience/climatology orbiter |
| 1990+ | Space Station | NASA | Manned | Major orbiting research facility |
| | ISO | ESA | Astronomy | Son of IRAS |
| | CXGT | Japan | Astronomy | X-ray telescope |
| | UVSAT | Japan | Astronomy | 60 cm UV telescope |
| | Starlab | NASA/AUS/CAN | Astronomy | 1 m UV/visible/IR |
| | AXAF | NASA | Astronomy | Advanced X-ray Astronomy facility |
| | CRAF | NASA | Exploration | Comet rendezvous/asteroid flyby |
| | LGO | NASA | Exploration | Lunar Geoscience Orbiter |
| | NEAR | NASA | Exploration | Near Earth Asteroid Rendezvous |
| | Titan probe | NASA/ESA | Exploration | Titan (Saturn moon) probe |
| | Galileo Saturn | NASA | Exploration | Saturn probe/orbiter |
| | Uranus probe | NASA | Exploration | Uranus atmosphere probe |
| | MSR | NASA | Exploration | Mars sample return |
| | Venus/asteroid lander | USSR | Exploration | Double mission to land on Venus and on the asteroid Vesta |

almost at the edge of the solar system. The spacecraft left Earth, together with its sister probe Voyager 1, in the summer of 1977 and sent back stunning pictures of Jupiter in 1979 and of Saturn in 1981. On Friday 24 January 1986 Voyager 2 flew past Uranus and its cameras returned the first-ever detailed pictures of that planet and its moons. The spacecraft was so far away that its radio signals took almost three hours to reach Earth – even though they travel at the speed of light.

As Voyager 2 paid our first-ever visit to Uranus it used the planet's powerful gravitational field to change direction and head on towards Neptune, another 1.5 light hours away at a distance of approximately 4.5 billion km (2.8 billion miles) from Earth. Voyager 2 is scheduled to arrive at Neptune on 24 August 1989 and, if all goes well, that encounter will be the culmination of a spectacularly successful 'Grand Tour' that was first planned by visionary NASA scientists back in the 1960s. By the end of this decade American spacecraft will have explored all but one of the eight planets that accompany Earth in orbit around the Sun. Only distant Pluto, a small cold outpost on the fringes of interstellar space, will remain unvisited.

This magnificent programme of planetary exploration, completed in a period of just 25 years, is arguably mankind's greatest technical achievement to date. Sadly, however, only one Voyager-class planetary mission has a firm place in NASA's plans for the foreseeable future. In May 1986, the Shuttle will carry Galileo into Earth orbit on the first stage of a two-year journey to Jupiter. Galileo is the most sophisticated planetary spacecraft ever built. (See Chapter Four.)

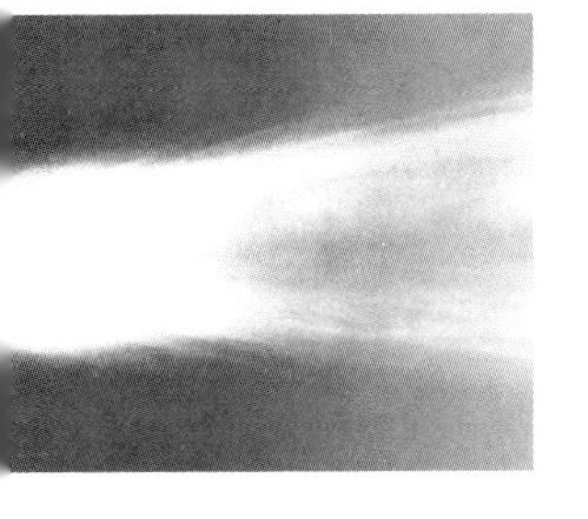

ABOVE Comet Halley's spectacular 1910 appearance.

RIGHT Giotto's adoration of the Magi, painted in the 14th century, includes a splendid and unmistakeable comet which inspired the European Space Agency to lend the artist's name to their spacecraft.

But when Voyager 2 makes its final encounter with Neptune in 1989, the first golden age of solar system exploration will come to an end. No more grand missions are in the pipeline, and it will be well into the next century before any spacecraft from Earth venture again to the outer solar system.

**Encounter with a comet**

There are, however, several exciting projects in preparation with targets closer to home. Following Voyager 2's encounter with Uranus in January 1986 there will be a blitz of activity in early March when Halley's Comet returns to our vicinity after an absence of 76 years. No fewer than five separate missions will be exploiting this rare opportunity to investigate a large comet at close quarters. A veritable fleet of spacecraft has been dispatched in an internationally coordinated effort to explore one of the most mysterious astronomical objects in the solar system.

The Soviet Union is sending two probes to Halley's Comet, Vega 1 and Vega 2, which are expected to pass respectively within about 10,000 and 3000 km (6200 and 1900 miles) of the central nucleus. Japan is sending two smaller spacecraft, MS-T5 – closest approach approximately 1,000,000 km (620,000 miles) – and the more sophisticated Planet A – closest approach 320,000 km (200,000 miles). Before the loss of *Challenger* NASA was planning to fly a special Shuttle/spacelab mission to observe the Comet but this flight was cancelled as a result of the disaster. The most spectacular visitor to Halley's Comet will, however, be a European spacecraft: Giotto.

Sometime during the night of 13 March 1986, Giotto will make a kamikaze dive straight into the dust cloud that surrounds the icy nucleus of Halley's Comet. The spacecraft and the comet will approach one another almost head on, each travelling in opposite directions around the Sun. The encounter will take place about 150 million km (95 million miles) from Earth after a journey across the solar system lasting six months. When they meet, Giotto's velocity relative to the comet will be a staggering 68 km (42.25 miles) per second – which is almost 250,000 km/h (155,000 mph) or 50 times faster than a rifle bullet. At this speed even the smallest grain of dust will have an enormous impact: a particle weighing just one-tenth of a gram becomes a deadly hazard capable of smashing its way through 8 cm or just over 3 in of solid aluminium!

No one really expects Giotto to survive the encounter, but to safeguard the precious scientific instruments as long as possible, the spacecraft has an ingenious two-part shield that provides maximum protection for minimum weight. The first part of the shield consists of a thin gold-plated aluminium sheet which can withstand small impacts, but that will allow larger particles to punch their way through. In the process of penetrating this first line of defence, the particle will vaporize and form a diffuse cloud of hot gas that spreads out before it reaches

BELOW The Giotto spacecraft to Halley's Comet will be the first European mission to leave the vicinity of Earth.

RIGHT The Ulysses spacecraft, attached to a Centaur inter-planetary booster, will be carried into orbit by the Shuttle. Following deployment the Centaur will fire, propelling Ulysses on a long arcing trajectory out to Jupiter and then back to swing over both poles of the sun.

the second part of the shield. This consists of a thicker 13.5 mm (0.53 in) Kevlar–foam sandwich which should be able to absorb the debris. The shield has been designed to withstand collisions with particles as large as 0.1 g. Whether Giotto will experience larger impacts depends on several unknowns, in particular the nature of Halley's Comet, which is what the mission is being sent to discover.

The climax of Giotto's mission to Halley's Comet will be an exciting time. A high definition colour camera will return the first-ever close-up pictures of the nucleus, which is believed to be a lump of ice several kilometres across, showing details down to a resolution of about 30 m (98 ft). In addition, a battery of scientific instruments will return a stream of unique and valuable data about the chemical composition of the comet's thin atmosphere. Comets are believed to have formed during the earliest stages of the solar system's evolution and astronomers all over the world would love to know more about them. In particular, Giotto's analysis of the various gases given off by the comet may help to decide between rival theories of how the solar system originally formed from a primordial cloud of hydrogen.

One of the most exciting things about Giotto's mission is that no one can predict what the spacecraft will find or what its pictures of the icy nucleus will be like. If the nucleus is surrounded by clouds of gas and vapour then the results may be disappointing. On the other hand there is a chance that the spacecraft may make a completely unexpected discovery. Happily, all the various countries that are sending probes to explore Halley's Comet have agreed to cooperate and to exchange information. The Russians, whose Vega spacecraft will encounter the comet several days before Giotto, will provide exact positional data so that last-minute adjustments can be made to Giotto's trajectory. This will ensure that Giotto passes as close as possible to the nucleus – as close as 500 km (310 miles), it is hoped. The Ides of March 1986 should be an interesting time.

**Exploring the solar system**

The two probes that the Soviet Union is sending to investigate Halley's Comet are involved in a mission which has even greater ambitions. On their way towards their encounter with the comet, both Vega spacecraft will first rendezvous with Venus. The main section of each vehicle will hurtle past the planet, taking high-quality pictures and making numerous scientific observations. Before the encounter, however, a small lander module will detach itself from each spacecraft and enter Venus' atmosphere. These landers will parachute to the surface where they will sample the soil and analyse its chemical composition. During their descent through the clouds, both the landers will release balloons equipped with a small gondola of scientific instruments. The 3.5 m (11 ft) balloons will float through the atmosphere radioing data back to the main Vega spacecraft as they speed off towards Halley's Comet thousands of kilometres above.

Soon after the excitement of Halley's Comet, two other important events in the field of solar system exploration are scheduled to occur: the Shuttle is due to launch Ulysses and Galileo, the first inter-planetary spacecraft to be dispatched in this way. All previous missions have left Earth riding atop conventional expendable rockets. Both missions, each requiring a dedicated Shuttle flight, will employ a specially modified Centaur upper-stage booster to blast out of low Earth orbit onward to their ultimate destination: the Sun in the case of Ulysses and Jupiter in the case of Galileo (see the next chapter).

Ulysses will be the second European spacecraft after Giotto to venture deep into the solar system. The probe is the remaining half of what was originally known as the international Solar Polar Mission. The name changed after NASA, in the midst of financial difficulties with the Shuttle, was forced to cancel its half of what was originally intended to be a two spacecraft mission. For a time, the United States' decision not to continue with the project severely damaged relations with the European Space Agency. It has undoubtedly been a significant factor in ESA's subsequent determination to pursue a largely independent space science policy.

Ulysses' purpose is to study the hitherto unexplored polar regions of the Sun. In order to achieve this objective, it will travel on a remarkable and paradoxical trajectory all the way out to Jupiter. Here it will use the enormous gravity of the largest planet in the solar system like a sling shot to swing back 'above' the north pole of the Sun. The journey out to Jupiter and back to the Sun, over which it should fly at an altitude of about 300 million km (185 million miles), will take almost four years. Ulysses will then swing down to make observations over the south solar pole, following which its nominal five-year mission will be completed.

Beyond 1986, only two further American planetary exploration missions are definitely scheduled. In 1988 NASA is due to launch the Venus Radar Mapper, which will be placed in a polar orbit to provide

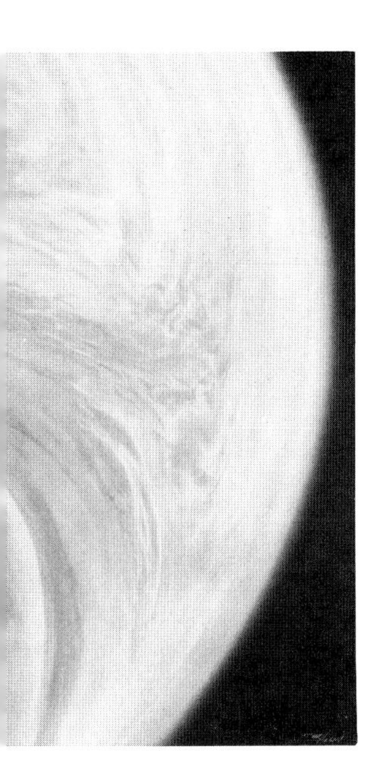

The Venus Radar Mapper is a low-cost mission that will orbit the planet for many months, returning high-resolution radar images of Venus's hidden surface.

comprehensive coverage of that planet. The spacecraft will be equipped with a sophisticated radar system capable of penetrating Venus's thick cloudy atmosphere and sending back to Earth images of the surface with a resolution perhaps as good as 500 m (1640 ft). The Venus Radar Mapper is in fact a scaled-down version of a more ambitious project, the Venus Orbiting Imaging Radar spacecraft, which was yet another victim of NASA's financial problems with the Shuttle.

The only other planned planetary mission, the Mars Geoscience/Climatology Orbiter, also reflects the spirit of economy that now pervades the field of space science. Scheduled for launch in August 1990, the spacecraft is due to arrive at Mars 12 months later. It will spend two Earth years (one Martian year) surveying the planet to provide detailed information about both the weather and the mineralogy on the surface. This data will be combined with high-resolution pictures from the immensely successful Viking Orbiters that operated during the mid-1970s to provide a comprehensive survey of Martian resources.

The Mars Geoscience/Climatology Orbiter will be a low-cost mission that makes use of an existing 'off-the-shelf' spacecraft originally designed for use in Earth orbit. NASA is looking at three candidates: the RCA defence meterological satellite, the Hughes HS-376 communications spacecraft and the TRW FleetSatCom vehicle. All would require considerable modification for such a new role, but NASA believes that this approach could save many tens of millions of dollars compared to designing a one-off custom-built spacecraft. Furthermore, the mission would employ existing instrumentation and communication hardware, which will also save considerable development costs.

Coincidentally, the Soviet Union is also planning a mission to Mars, currently scheduled for launch in 1988. If the rumours are correct, the mission will involve a pair of spacecraft each equipped with a small detachable probe designed to land on one of the two small Martian moons, Phobos and Demos. This will be the first Soviet flight to Mars since the early 1970s and yet the project will be far more ambitious than NASA's Mars Orbiter that is due to arrive about two years later.

Two other Soviet planetary missions are thought to be in the pipeline. The first is a lunar polar orbiter designed to provide geochemical mapping of the Moon's entire surface. It would be launched in 1989 or 1990 and it appears to be very similar in concept to a mission NASA is also considering (see below). The final planetary project that Russian space scientists are believed to be developing would involve a return to Venus combined with a landing on an asteroid. This mission, designated 'Vesta', would once again involve a pair of spacecraft. One vehicle would deliver a lander to descend through the atmosphere of Venus and then continue on to rendezvous with an asteroid. The other vehicle would head directly for a different asteroid and release a lander to sample its surface. Apparently this mission has yet to win official approval, but the aim is for launch to take place in 1991.

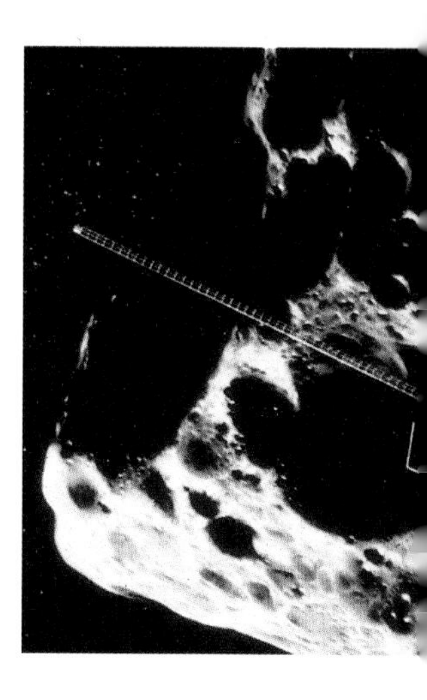

**Astronomy in space**

Following manned research and planetary exploration, the third major type of space science operation is astronomical observation. Telescopes and other instruments mounted on satellites in Earth orbit have a view of the heavens superior to that from any ground-based observatory. On Earth, the atmosphere blocks most celestial radiation and even optical wavelengths, which can travel through air, are adversely affected by water vapour and dust particles. Astronomers are free to observe the sky using all the gamma rays, X-rays, ultraviolet and infrared wavelengths that never make it to the surface. And they can also make more traditional observations in the visible spectrum with unprecedented clarity and sensitivity.

The number one forthcoming astronomical highlight in space, and indeed the number one development in astronomy in general for years to come, is the scheduled launch of the Hubble Space Telescope in the autumn of 1986. (As mentioned earlier, 1986 looks like being a golden year for space science!) The Space Telescope will do for present-day astronomers what Galileo's telescope did for the human eye. It will give an order-of-magnitude improvement in both resolving power and ability to detect faint objects and its arrival is likely to provoke a cascade of major astronomical discoveries during the remainder of the century. The Space Telescope is discussed in detail in Chapter Five.

Although the Space Telescope is by far the most expensive and important forthcoming development in space astronomy, it is by no means the only project in the pipeline. At least eight other astronomical spacecraft are under construction for launch during the remainder of the 1980s, and several ambitious missions have been proposed for the early 1990s. A significant trend that can be discerned in future plans for space astronomy is the growing importance of international collaboration and the end of the United States' traditional dominance in this area of space science. Even NASA's Space Telescope, which at $1200 million has cost more than all remaining astronomical missions for the rest of the decade, is a joint programme with the European Space Agency. ESA is building several key items of hardware and meeting 15 per cent of the cost of ST's development.

America has joint scientific ventures planned with West Germany, Australia and Canada; and NASA is not alone in seeking others with whom to share the expense of space astronomy. The Soviet Union is collaborating with France in developing an orbiting telescope that will use detectors cooled by cryogenic techniques to −230°C for observing the sky at sub-millimetre wavelenths. Japan is building a large X-ray satellite, called Astro-C, in conjunction with Great Britain. Both these spacecraft are due to be launched in 1987.

**An uncertain future**

Considering the prospects for space science in the 1990s, little is certain

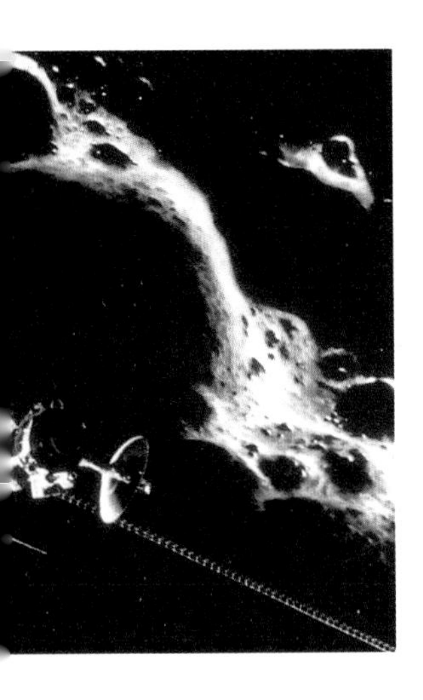

NEAR is a mission for which NASA is still seeking funding. The acronym stands for Near Earth Asteroid Rendezvous and the spacecraft could help to discover whether there are economically attractive deposits of precious metals waiting to be recovered from the vast lumps of rock that occasionally pass by our planet.

except that Space Station will become the focus of manned American operations. NASA maintains that the huge expense of developing and operating the Station will not affect the future of unmanned projects. But many space scientists, who have bitter memories of similar assurances in the early 1970s regarding the Shuttle, are sceptical.

Several missions currently under consideration will follow the example set by the Mars Geoscience/Climatology Orbiter in using an existing standard planetary observer spacecraft. One of the more ambitious proposals is called NEAR, standing for Near Earth Asteroid Rendezvous. This mission would provide the first detailed information about the composition of the asteroids that occasionally approach the Earth and may one day provide a valuable source of raw materials. Possible targets are the asteroids Eros in 1990, Anza in 1991, Aneros in 1992 or 1993, or Ivar in 1995. Another proposed planetary observer class mission is the Lunar Geoscience Orbiter. LGO would be the first US mission to the Moon since Apollo and its information could pave the way for construction of a manned lunar base early next century (see Chapter Nine).

For operations further afield, NASA is planning to develop a more sophisticated class of spacecraft that will be known as Mariner Mark 2. These standardized vehicles will be similar to the Voyager and Galileo designs, and once again they will make use of existing hardware wherever possible. Although this approach means that planetary exploration will no longer be exploiting state-of-the-art technology, NASA believes it will be able to afford a greater number of missions overall.

Definite missions for Mariner Mark 2 class spacecraft have yet to be confirmed and the first launches will not occur before the early 1990s. Likely objectives include a comet rendezvous/asteroid fly-by, a Titan probe (probably a joint NASA/ESA project), a Saturn probe/orbiter that would follow on from Galileo, and even perhaps a Uranus probe. Looking still further into an uncertain future, NASA is proposing some ambitious unmanned missions near the turn of the century. Undoubtedly the most exciting is the Mars Sample Return mission. This would involve construction of a large multi-stage spacecraft in Earth orbit and would ultimately make use of Space Station as a receiving laboratory for Martian soil.

However, none of the longer-term space science missions mentioned above is definitely funded. Many, like the Saturn probe/orbiter and the Advanced X-ray Astronomy Facility, have already been postponed for years since they were originally proposed. There are no truly outstanding projects in astronomy or planetary exploration firmly scheduled for the 1990s and, with such long lead times involved, a barren period seems inevitable. All the exciting highlights of 1986 are the fruit of decisions, commitments and investments made during the last decade. It is sad to think that, for the present at least, no similar seeds are being sown.

CHAPTER FOUR

# Galileo

*On the seventh day of January in the present year 1610, at the first hour of the night, when I was viewing the heavenly bodies with a telescope, Jupiter presented itself to me; and because I had prepared a very excellent instrument for myself, I perceived (as I had not before, on account of the weakness of my previous telescope) that beside the planet there were three starlets, small indeed, but very bright.*

Galileo Galilei, astronomer and scientist

Thus the great Italian astronomer Galileo Galilei recorded his discovery of the moons of Jupiter. It was a momentous observation for it strongly supported the Copernican picture of the solar system according to which the planets, including the Earth, orbit the Sun just as the Galilean satellites in turn orbit Jupiter. The Church hounded Galileo mercilessly for associating himself with this new scientific view that threatened to undermine the traditional, religious view that set the Earth at the centre of the universe. But in the long run Galileo's belief has become the accepted truth and more than three centuries after his death, the great astronomer is about to be honoured in a way that would have amazed and delighted him: Galileo is going to Jupiter.

Galileo Galilei (1564-1642) Italian astronomer, physician and pioneer of the experimental method.

## A mission to Jupiter

Galileo's name has been given to NASA's next great planetary exploration mission. To date four probes have visited Jupiter: Pioneer 10 and 11 in 1973 and 1974, and Voyager 1 and 2 in 1979. But all these probes were on fly-by missions, encountering Jupiter only briefly as they hurtled past the planet on their way through the solar system. Spectacular as the results were, they have provided only a brief and tantalizing glimpse of Jupiter and its family of moons. Galileo is designed to consolidate our exploration of Jupiter by making an extended and detailed survey of the entire Jovian system. The spacecraft will place itself in orbit around the planet, sending back a treasure trove of data over a period of at least two years. Galileo will also send a separate probe vehicle actually down into Jupiter's atmosphere – providing in just a few minutes of observation more information about the nature of the planet than has been gathered in all the years since Galileo first turned his telescope to the sky.

The Galileo mission's numerous scientific objectives fall into three distinct categories. Firstly, the spacecraft will investigate the structure and composition of Jupiter's atmosphere. Secondly, Galileo will undertake detailed exploration of the planet's four principal moons: Io, Europa, Callisto, Ganymede. The spacecraft will not only return extremely detailed pictures (with a resolution hundreds of times better than Voyager) but will also make spectrographic measurements to reveal the chemical make-up of Jupiter's satellites. Thirdly, the mission has been designed to investigate all aspects of the planet's magnetosphere, the vast dynamic region of space that falls under the influence of Jupiter's powerful magnetic field.

Galileo achieves this wide variety of objectives in spectacular fashion. Both the spacecraft hardware and the mission plan have been designed with fantastic care and attention. Galileo represents the current state of the art in space technology, drawing on the skills of scientists and engineers from a hundred different fields and costing almost a billion dollars. Although Galileo is following in the path of the two Voyager spacecraft, the new mission will be far more than simply 'Voyager 3'.

The story begins when the Shuttle carries Galileo up to a low Earth orbit. Once there, however, there is still a long way to go and so a high-energy Centaur upper stage is used to boost the spacecraft out of Earth's gravity field on a trajectory that two years later will bring it to Jupiter. This Centaur booster is fuelled with liquid oxygen and liquid hydrogen and it is a design that has produced over 40 consecutive successful launches. Special modifications have been made to fit the cargo bay of the Shuttle, however, and there is some nervousness among the Galileo team as to whether this untried technique of planetary launch will work first time. In fact another spacecraft (the European Ulysses mission to orbit the Sun) is also scheduled to launch with an identical Shuttle–Centaur combination at about the same time. There is apparently an unofficial competition between the two teams to be second!

Once safely in Earth orbit explosive bolts will fire and steel springs will gently push the Centaur, with Galileo attached, away from the Shuttle. The main Centaur engine then ignites and burns for about ten minutes to insert Galileo in the appropriate Earth–Jupiter trajectory. The Centaur's final action is to stabilize the vehicle as it deploys its large 48 m (15.75 ft) antenna and 10.9 m (35.75 ft) instrument boom, and then gently to spin the spacecraft before release.

This 3.15 rpm spin is one of the main features governing Galileo's design. It gives the spacecraft its stability and allows certain scientific instruments continuously to sample the space environment in all directions. But the spin is less than ideal for other devices: particularly remote-sensing imaging instruments, which need to be pointed accurately in three-dimensional space. So, to accommodate cameras, telescopes, spectrographs and tracking antennae, a section of the Galileo spacecraft is 'de-spun' – that is, rotated in the opposite direction with respect to the spinning main body of the vehicle so as to remain effectively motionless relative to the stars. This is the first time that such a complex design has been used on a planetary vehicle and it has given Galileo many advantages as well as the designers many headaches.

During the two-year cruise to Jupiter relatively little activity takes place on board the spacecraft. Instruments will be callibrated and a limited amount of data may be gathered to investigate the interplanetary environment. At the time of writing, NASA was trying to decide whether Galileo should be diverted *en route* to encounter one of the large asteroids that orbit the Sun between Mars and Jupiter. Amphitrite, an irregular lump of rock approximately 200 km (125 miles) across, would be an ideal target. The scientific reward, including images of the asteroid with a resolution better than 200 m (655 ft), makes this mission option tempting. But the decision to rendezvous with Amphitrite has two serious penalties: Galileo's arrival at Jupiter would be delayed by three months and close encounter with a never-before explored asteroid inevitably carries risks that could compromise the mission's true purpose. It is likely that a final decision will only be made after launch, when the status of the spacecraft has been checked.

Two other important events, in addition to the possible asteroid fly-by, stand out during the Earth-to-Jupiter cruise phase.

About eight months after departure, Galileo will fire its thruster engines to move out of the plane of Earth's orbit and into the plane of Jupiter's orbit, which is slightly inclined with respect to the solar system as a whole. Then, 150 days before arrival at Jupiter, Galileo will split into two parts: the main Orbiter spacecraft, which will tour the Jovian system for at least two years, and the Probe, which will actually enter the atmosphere of the planet.

In this montage of Voyager 1 images, Jupiter is in the top right of frame. Io is towards the upper left, Europa is in the centre, Ganymede is to the lower left and Callisto is at the bottom right.

Galileo will arrive at Jupiter separated into two spacecraft: the smaller probe will plunge into the planet's atmosphere (left); while the main orbiter will visit each of Jupiter's moons, beginning with sulphurous, volcanic Io (right). (The spherical profile of an explosive eruption can be seen rising from the surface of Io.)

Just before separation an explosive guillotine severs the cable linking the two Galileo vehicles and from that moment on, no instructions can reach the Probe. But even if it could receive messages from Earth or from the Orbiter, the Probe is incapable of making any alteration to its flight path. At release its trajectory must be exactly on target for an encounter with Jupiter at a precisely predetermined angle, velocity and location on the planet 150 days into the future! Following separation the Orbiter briefly fires its thrusters to back off from the Probe, ensuring that it will safely miss the atmosphere and be in a position to relay the Probe's signals back to Earth.

**Probing the atmosphere**

The factors determining the Probe's entry into Jupiter's atmosphere are critical. Accelerated by the massive gravity of the planet, the Probe will be travelling at about 180,000 km/h (112,000 mph) relative to the atmosphere. The mission designers have deliberately selected a target on the side of the planet that is turning away from the Probe's direction of travel – thus making use of Jupiter's considerable rate of rotation to reduce the spacecraft's relative speed and minimize the impact of entry into the atmosphere. Even so, the probe will be travelling more than four times faster than the Apollo astronauts did on their return from the Moon. No man-made object has ever survived atmospheric entry at such speeds. About two-thirds of the Probe's 335 kg (740 lb) mass is devoted to the heatshield and it is anticipated that over half of the Probe's total weight at launch will be burned away during the first few minutes of entry. Crushing deceleration forces up to 345 times Earth gravity will briefly act on the spacecraft as it begins to encounter the denser parts of Jupiter's atmosphere.

But if all goes well the computers, transmitter and scientific instruments on board the Probe will survive safe inside the heatshield, remaining more or less at room temperature while only a few centimetres away the outer surface will reach 14,000°C. Sensing the drop in g forces once the vehicle has been slowed to a relatively low speed by Jupiter's atmosphere, the on-board computers will command an explosive mortar to fire, ejecting a small pilot parachute that pulls away the 'lid' of the Probe and allows the main parachute to be deployed. This 2.5 m (8.2 ft) parachute will slow the descent module to a mere 150 km/h (93 mph) per hour, allowing the remains of the heatshield to drop away, so exposing the Probe's precious cargo of scientific instruments. Then, for between 40 and 60 minutes, the Probe will radio back unique and priceless data as it descends down through Jupiter's cloud layers, becoming the first man-made object to enter the planet's atmosphere.

The crucial hours when Galileo arrives at Jupiter will be an incredibly busy time for mission controllers back on Earth. Not only will they be monitoring the data returned by the Probe, but they will also be supervising some critical manoeuvres on the part of the main Orbiter spacecraft which must insert itself into a carefully determined Jupiter orbit. In a short space of time the Orbiter has no fewer than three major tasks to perform.

First, about four hours before the Probe hits the atmosphere, the Orbiter encounters Io, passing within 1000 km (620 miles) of Jupiter's spectacular volcanic moon. This rendezvous serves the twin purposes of providing very high resolution photographic and spectrographic coverage of the satellite, and, by using Io's gravity to deflect its trajectory, the Orbiter will substantially reduce the size of the burn it must subsequently make to achieve Jupiter orbit. Secondly, as soon as the Orbiter has returned its images of Io, it must place itself into the correct orientation to relay to Earth the data from the Probe as it plunges into Jupiter's atmosphere below. For this purpose the Orbiter has a special steerable antenna on the de-spun section that will track the Probe as the spacecraft flies overhead. Finally, as soon as the data link with Probe is lost (either because the radio signals can no longer penetrate the clouds or because the Probe's electronics have been crushed by the growing atmospheric pressure) the Orbiter must fire its main engine to become the first artificial satellite of Jupiter.

### Celestial billiards

If Galileo is not diverted to fly-by an asteroid, then the spacecraft should reach Jupiter about 800 days after leaving Earth. Although the day of arrival will be the highlight of the mission, with Io encounter and Probe entry occurring just a few hours apart, it is really only the beginning of the story. Over the following 20 months the Galileo Orbiter will perform a remarkable feat of celestial navigation. Using a technique known as gravity assist, the Orbiter will undertake a comprehensive tour of the Jovian system, visiting in turn each of Jupiter's principal moons and sweeping through all the different regions of the planet's magnetosphere.

To accomplish its complex tour of Jupiter's moons, Galileo exploits the gravitational influence of each body. An encounter deflects the spacecraft's path by a precisely calculated amount onwards to its next appointment.

## Gravity Assist

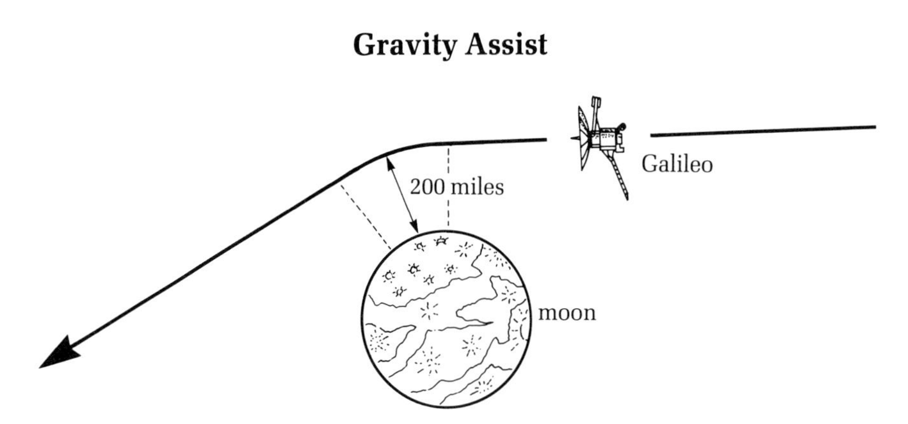


The gravity assist technique has sometimes been described as celestial billiards since it involves targeting the spacecraft like a cue ball bouncing repeatedly off the side cushions of a billiard table. In the case of Galileo, the cushions are the moons of Jupiter and there are no fewer than 12 bounces in the shot! The idea is to pass in turn very close to each moon on the tour, so that each time the satellite's gravity precisely deflects the spacecraft on towards its next encounter. Using this technique the Orbiter can achieve up to 15 fly-bys of Jupiter's moons during its 20-month tour. Five of these encounters will be at altitudes below 1000 km (620 miles), including a spectacular 200 km (125 miles) swoop over ice-covered Europa which promises to yield pictures of that satellite with a resolution as good as 4 m (13 ft).

The great advantage of the gravity assist technique is that it allows Galileo to make its tour of Jupiter's moons using very little fuel. Essentially the engines will be used only to correct small navigational errors and to keep the spacecraft's antenna accurately pointed towards Earth. If gravity assist were not employed, then to achieve the same tour, Galileo would have to carry about 60 times more fuel than it otherwise does, which would make the spacecraft far too heavy to be launched from Earth. Without playing celestial billiards with gravity assist there could be no Galileo mission.

Although billiards provides a graphic analogy for Galileo's style of navigation, it is misleading both in terms of the precision required and in terms of the complexity of the problem. Billiards takes place on a flat table with fixed targets. Jupiter and its moons follow highly complex paths in three-dimensional space and, what is worse, they are continuously moving relative to the Earth, to the Sun and to one another. Galileo must be targeted at an empty point in space, many millions of kilometres away and hundreds of days in the future. The Orbiter must arrive at exactly the point in space and time when the gravity of Jupiter and its moons will have precisely the desired effect of hurling the spacecraft on to its next encounter, where all the same calculations must be repeated, and so on for the entire 20-month tour!

### Touring the moons

The design of the Galileo Orbiter's mission is an incredibly skilled process that must reconcile a great many different factors. Needless to say the entire operation would be impossible without powerful high-speed computers and extremely sophisticated software. The aim, of course, is to extract the maximum amount of scientific information about the Jovian system during the 20-month working life of the Orbiter.

But the navigators' task is not simply to visit as many satellites as possible. They must ensure, for example, that each encounter crosses the sunlit face of a moon (otherwise the cameras would be useless); and when the Orbiter returns to a satellite the tour designers must try to ensure that the spacecraft flies over a new region, different from previous visits. Encounters must take place at a time when conditions are favourable for communication with Earth. And at all times spacecraft safety is paramount. Cumulative radiation exposure is carefully estimated so that Galileo only exceeds its total budgeted allowance of 150 krads right at the end of the tour: as far as possible avoiding tempting but dangerous targets until towards the conclusion of its mission.

With all these constraints in mind, Galileo's mission designers have planned a tour for the Orbiter that should meet all the major scientific objectives while minimizing risks and uncertainties. During the 20 months after arrival at Jupiter, the spacecraft will follow a total of 12 long looping orbits through the Jovian system, experiencing a close encounter with at least one moon on each circuit. The orientation of these eccentric orbits will gradually swing around during the course of the tour (a change deliberately engineered using gravity assist) so that in

**Orbital tour**

| *Date (D/M/Y)* | *Encounter (*)* | *Orbit (*)* | *Sattelite (name)* | *Altitude (km)* | *Resolution (BEST/m)* | *Radiation (tour total/krad)* |
|---|---|---|---|---|---|---|
| 25/08/88 | 1 | arrival | Io | 1,000 | 20 | 50.0 |
| 18/03/89 | 2 | 1 | Ganymede | 834 | 17 | 52.0 |
| 21/05/89 | 3 | 2 | Ganymede | 1,146 | 23 | 54.0 |
| 22/07/89 | 4 | 3 | Callisto | 494 | 10 | 63.5 |
| 01/09/89 | 5 | 4 | Europa | 1,400 | 28 | 73.0 |
| 10/10/89 | 6 | 5 | Europa | 200 | 4 | 82.5 |
| 09/11/89 | 7 | 6 | Callisto | 6,346 | 127 | 92.5 |
| 02/12/89 | 8 | 7 | Europa | 25,000 | 500 | 109.0 |
| 03/12/89 | 9 | 7 | Ganymede | 3,258 | 65 | 109.0 |
| 21/12/89 | 10 | 8 | Ganymede | 1,722 | 34 | 123.5 |
| 17/01/90 | 11 | 9 | Europa | 472 | 9 | 137.5 |
| 22/02/90 | 12 | 10 | Callisto | 40,630 | 813 | 147.5 |
| 23/02/90 | 13 | 10 | Ganymede | 3,680 | 74 | 147.5 |
| 13/04/90 | 14 | 11 | Callisto | 277 | 6 | 150.0 |
| 14/04/90 | 15 | 11 | Ganymede | 45,955 | 919 | 150.0+ |
| 29/05/90 | – | 12 | (tail petal of Jupiter's magnetosphere) | | | |

A timetable of Galileo's tour around the moons of Jupiter assuming arrival on 25 August 1988.

Galileo nears completion at JPL's Spacecraft Assembly Facility in Pasadena, California. For a sense of scale, note the technician standing towards the left end of the long magnetometer boom.

between visiting the moons, the Orbiter will fly through a cross section of Jupiter's magnetosphere.

Galileo will visit all the Jovian moons at least twice except for Io which it encounters just once, right at the beginning of the tour. Io lies close to Jupiter itself towards the region of greatest radiation. The Orbiter will receive a third of its total budgeted radiation exposure in the first few hours after arrival and it will not venture so close to the planet again. At the very end of the tour Galileo will fly low over Callisto and use the moon's gravity to swing right out into the unexplored region of space that lies behind Jupiter as the planet orbits the Sun. Here in the wake of the magnetosphere, Galileo will make its final planned observations, investigating the interaction of Jupiter's magnetic field with the solar wind.

## The spacecraft

Galileo is technically the most advanced spacecraft ever sent to explore the solar system. Its complexity and sophistication stem from the requirement simultaneously to fall within the maximum acceptable launch weight, to carry out the most extensive scientific mission possible, and above all to accomplish its chosen tasks with a very high probability of success. Galileo must be reliable, resilient and resistant to failure. It would be disastrous if a billion dollars, not to mention years of effort by some of the world's greatest scientists and engineers, were wasted because of the failure of some minor component.

The spacecraft is thus the product of incredibly careful design. Redundancy is built into all the main systems so that no single failure will jeopardize the mission as a whole. The on-board computers that control every event during the flight have been given an unprecedented degree of autonomy so that Galileo will be able to safeguard itself in the event of a temporary loss of communication. This autonomy also allows the spacecraft to respond immediately to changing conditions, rather than waiting for instructions to travel the enormous distance from Earth.

The two main components of Galileo, the Probe and the Orbiter, have a combined mass of 2550 kg (5620 lb) at departure from Earth; the Probe is the smaller of the two weighing 345 kg (760 lb). There are a total of 14 scientific instruments carried on the mission: 6 on the Probe and 8 on the Orbiter. In addition, the spacecraft's radio signals can be tuned to yield scientific information, for example when Galileo passes behind Jupiter so that its signal beam progressively sections the atmosphere.

The instrument that will interest most people back on Earth is naturally the camera which, for the first time, will employ a solid-state imaging device rather than the conventional television tubes used on previous planetary spacecraft. The optics, however, are identical to the Voyager 1500 mm focal-length telescope, and in fact Galileo is economizing by using an old Voyager spare. However, the quality of the pictures returned by the new camera promise to be superb: all the satellites will be mapped with a resolution 20 to 300 times better than the images from Voyager. In part this improvement is due to the 800×800 resolution CCD imaging device which is 100 times more sensitive than Voyager's vidicon camera. Another major factor will be the greatly reduced encounter distances. As well as providing 20 m (66 ft) resolution coverage of all the Jovian moons, Galileo's camera will also regularly be turned to observe Jupiter itself. During the 20-month tour the Orbiter should build up a uniquely detailed record of the planet's weather over an extended period of time.

**Science payload**

| *Instrument* | *Location* | *Objectives* |
|---|---|---|
| Solid state imaging | Orbiter (de-spun) | Map Jupiter's moons with a resolution of 1 km or better. Provide weather pictures of Jupiter during the mission. |
| Near infrared mapping spectrometer | Orbiter (de-spun) | Map Jupiter and its satellites at infrared wavelengths to investigate temperature and mineralogy. |
| Ultraviolet spectrometer | Orbiter (de-spun) | Study composition and structure of Jupiter's upper atmosphere and investigate Io's sulphurous wake. |
| Photopolarimeter radiometer | Orbiter (de-spun) | Measure temperature profiles and energy balance of Jupiter's atmosphere and obtain thermal data on moons. |
| Magnetometer | Orbiter (spun) | Investigate magnetic fields in the Jovian system. Discover whether any satellites have their own field. |
| Plasma detector | Orbiter (spun) | Provide information about clouds of ionized gas and low-energy particles in the magnetosphere. |
| Energetic particle detector | Orbiter (spun) | Measure the composition, distribution and energy of particles trapped in Jupiter's magnetic field. |
| Dust detector | Orbiter (spun) | Determine size, speed and charge of small particles, such as micrometeorites, in the Jovian system. |
| Atmospheric structure instrument | Probe | Record the temperature, density, pressure and molecular weight of Jupiter's atmosphere during probe descent. |
| Neutral mass spectrometer | Probe | Analyse the chemical composition and the proportion of gases at different levels of Jupiter's atmosphere. |
| Helium abundance detector | Probe | Measure with high accuracy the ratio of hydrogen to helium in Jupiter's atmosphere. |
| Nephelometer | Probe | Determine the location and composition of cloud layers at different heights in Jupiter's atmosphere. |
| Net-flux radiometer | Probe | Investigate the ambient thermal and solar energy at different altitudes during the probe's descent. |
| Lightning and radio emission detector | Probe | Verify the existence of lightning in Jupiter's clouds by detecting flashes and radio crackles. This instrument will also investigate Jupiter's inner radiation belts. |

LEFT The Galileo spacecraft has a payload of fourteen valuable scientific experiments.

OPPOSITE Anatomy of Galileo. The atmospheric probe is illustrated attached to the main spacecraft and in an expanded view above.

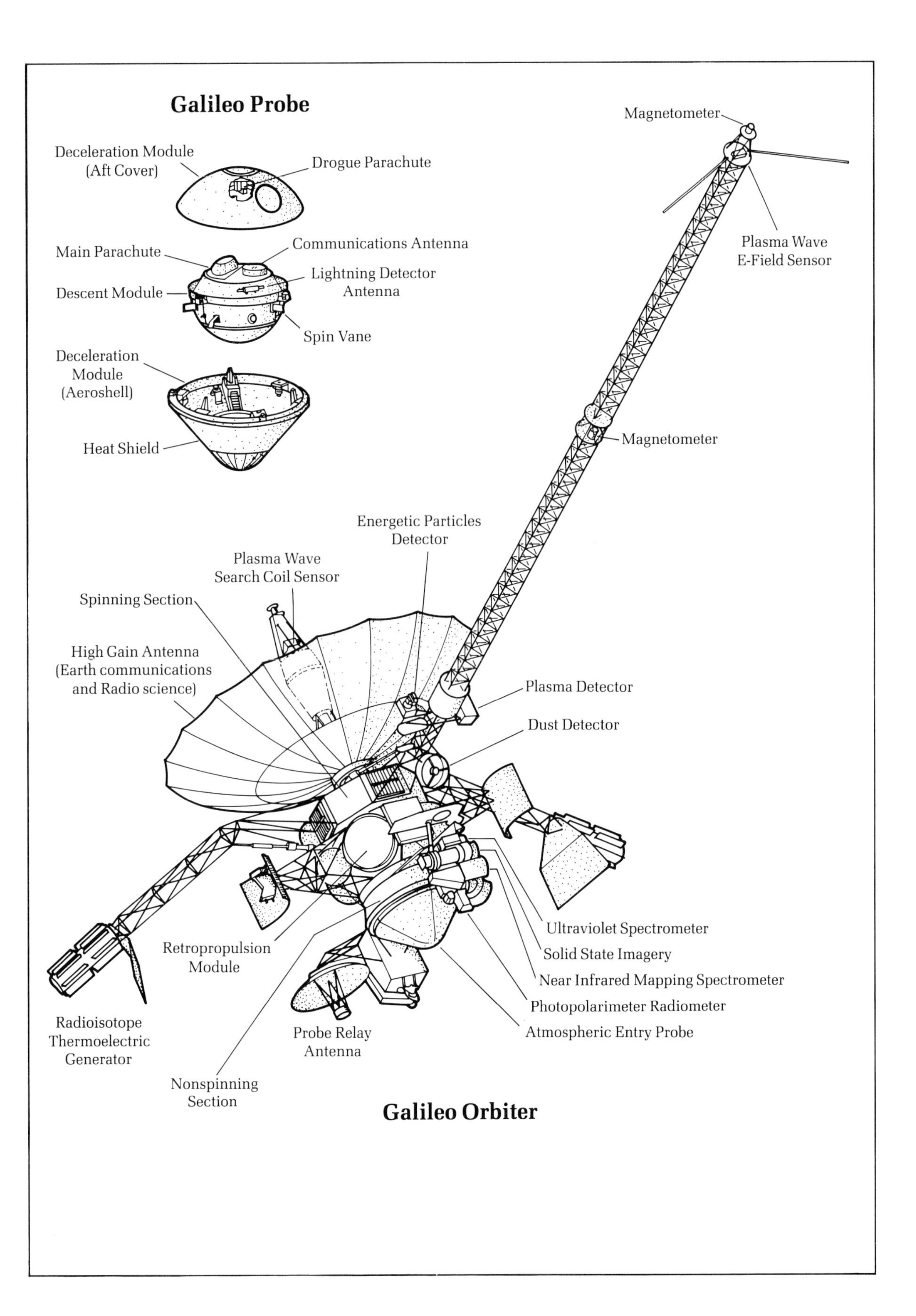

Galileo Probe
Deceleration Module (Aft Cover)
Drogue Parachute
Main Parachute
Communications Antenna
Lightning Detector Antenna
Descent Module
Spin Vane
Deceleration Module (Aeroshell)
Heat Shield
Magnetometer
Plasma Wave E-Field Sensor
Magnetometer
Energetic Particles Detector
Plasma Wave Search Coil Sensor
Spinning Section
High Gain Antenna (Earth communications and Radio science)
Plasma Detector
Dust Detector
Ultraviolet Spectrometer
Solid State Imagery
Near Infrared Mapping Spectrometer
Photopolarimeter Radiometer
Atmospheric Entry Probe
Retropropulsion Module
Radioisotope Thermoelectric Generator
Probe Relay Antenna
Nonspinning Section
Galileo Orbiter

To return all the data from its payload of scientific instruments, Galileo has the most powerful communications system ever installed on an interplanetary spacecraft. All components in the telecommunications system are completely duplicated since a failure here would be catastrophic for the mission. The large 4.8 m (15.75 ft) furlable antenna is the most prominent feature on the vehicle and it is capable of transmitting data back to Earth at the rate of 134,000 bits per second: that is fast enough to send all of the text in this book in about 20 seconds! The Orbiter will be able to send images from its camera back to mission control as quickly as they can be recorded and processed, which is approximately one picture every two seconds. It is astonishing to think that Galileo will send us these high-resolution snapshots from a distance of several hundred million kilometres, using little more power than it takes to run a domestic light bulb.

Electrical power for Galileo comes primarily from a pair of nuclear generators that are kept well away from the spacecraft at the end of extendable booms. They will supply an average of around 500 watts throughout the four years of the mission.

The spacecraft's propulsion system is being built by West Germany and it consists of a large module that is located on the main spinning section of the Orbiter. At the beginning of the mission Galileo will be carrying 932 kg (2055 lb) of fuel, which is well over a third of the vehicle's total mass. There are 12 small 10 newton thrusters and one large 400 newton engine, which is used primarily for the burn that will put the spacecraft into Jupiter orbit. The large engine can only be used after the Probe has separated from the Orbiter since its exhaust nozzle is not exposed until after that event. Aside from the few major burns to adjust the trajectory, the main function of the propulsion system is to keep Galileo properly oriented in space: both so that its antenna is pointed accurately at Earth and so that the camera and other instruments can be aimed towards their target.

The final major system on board Galileo is difficult to see since it is distributed throughout the spacecraft. The Command and Data System is perhaps the most significant technical innovation in the project, making use of the most up-to-date electronic hardware and computer design techniques. Never before has so much flexibility and sophistication been built into a spacecraft. Galileo is the most intelligent robot explorer that the human race has ever sent to investigate another world.

Unlike previous planetary spacecraft, there is no single master computer on board Galileo. Instead almost every device has its own microprocessor and the network is linked by a high-speed data bus. Failure of any one device should never prevent other systems from operating. Ironically, one of the most difficult problems that the engineers faced in implementing this sophisticated computer architecture was a mechanical one: how to provide a high-capacity data link between the spinning and the de-spun sides of the Orbiter. Slip rings and roller bearings were

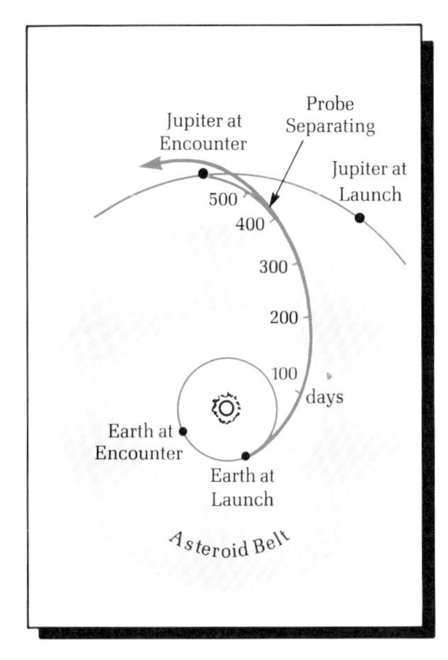


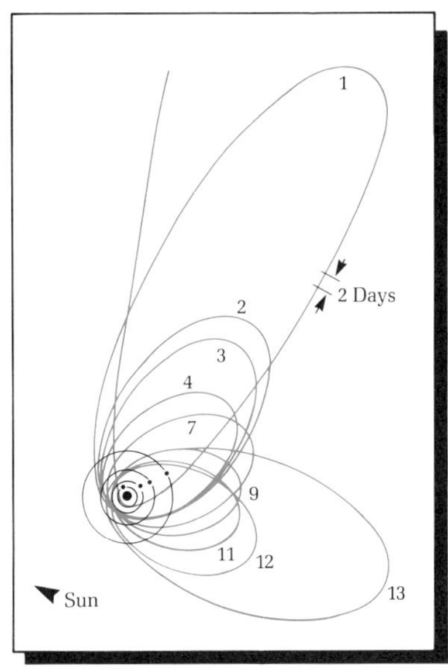


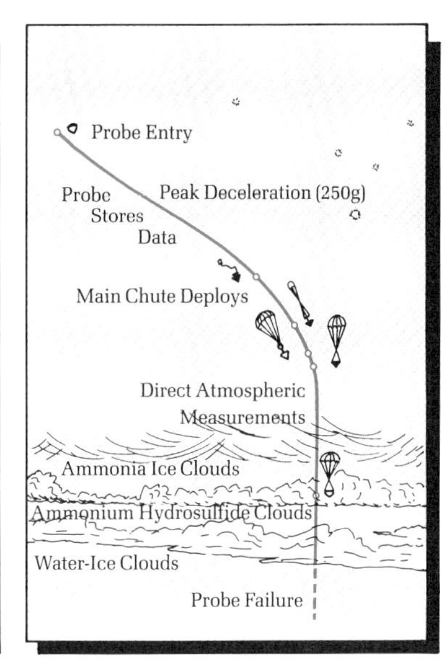


Three stages of the Galileo mission: the journey from Earth to Jupiter (left), the Orbiter's complex tour of the Jovian moons (centre), and the Probe's short-lived descent into Jupiter's atmosphere (right).

tried but they were either too noisy or else too unreliable to last for the duration of the mission. In the end rotary transformers were used.

Another problem relating to the electronic systems concerns the need to protect Galileo's vital computers from the harsh levels of radiation that the spacecraft will encounter at Jupiter. Modern integrated circuits can easily be affected by radiation: in high fields they begin to make mistakes. A single binary '1' accidentally changed to a '0' could destroy the entire mission – if it caused the main engine to fire incorrectly for example. To minimize this danger, Galileo's designers have turned to the military and made use of special hardened components developed to survive the similar conditons that would be found during a nuclear war. A rare case of swords being turned to ploughshares perhaps!

Galileo has been designed to function for at least 20 months after arriving at Jupiter. At the end of that time, if the mission has gone exactly according to plan, the Orbiter should be left with a few last drops of fuel in its tanks ('riding on fumes' as they say) and the spacecraft's electronics should have received the maximum safe dose of radiation. Galileo will probably end its useful life when the thrusters can no longer keep the antenna accurately pointed towards Earth – even though its computers and power packs will continue working for years.

But the spacecraft that NASA's Jet Propulsion Laboratory send out into the solar system have a habit of surviving longer than they should. The specifications for the Viking Landers that touched down on Mars called for the vehicles to operate for 90 days after arrival. In reality the last message from Viking 1 was received in November 1982, more than six years after it landed on Mars! There is a reasonable likelihood that a similar extension will be granted to Galileo and the project scientists are keeping an open mind as to what the Orbiter might do after its mission has formally been completed. One option is to send the spacecraft into orbit around Io for a final close-up inspection of that moon's magnificent volcanoes. The high levels of radiation there mean that Galileo would not survive for long, but it would certainly be a glorious end.

CHAPTER FIVE

# Space Telescope

*If the doors of perception were cleansed everything would appear to man as it is, infinite.*

William Blake, from The Marriage of Heaven and Hell

The star-filled sky on a clear night is one of the most magnificent sights in all of nature. Yet on Earth we see the universe as if looking through a small and grubby windowpane. Even from the highest mountain top, the atmosphere distorts and blocks our view, blurring detail and preventing all but a small fraction of starlight from reaching the ground. The telescopes in the world's largest observatories have reached the limit of resolution that is possible when viewing stars through the atmosphere and astronomers have long dreamed of escaping this frustrating restriction. In 1986 their dream will become reality when the Shuttle launches the Hubble Space Telescope.

## A revolution in astronomy

The vital statistics of the new telescope, known by those in the field as ST, are impressive. Weighting 11,000 kg (24,250 lb) and costing (according to the latest estimate) $1.2 billion, the Space Telescope is the largest and the most expensive scientific spacecraft ever built. It is also likely to be the most significant, revolutionizing every area of astronomy and providing answers to a library of mind-bending questions. Among the more exciting possibilities, ST should discover the age of the universe, determine its likely fate, decide between rival versions of the Big Bang theory, confirm the existence of black holes and even, perhaps, spot planets in orbit around other stars.

In purely numerical terms, ST will be at least as big an astronomical leap as occurred when Galileo first used a telescope rather than the naked eye to look at stars. In its most sensitive mode, ST will be capable of resolving details as fine as 0.007 arc seconds (there are 60 arc seconds in one arc minute and 60 arc minutes in one degree). This resolution is good enough to recognize a human face from a distance of more than 100 km (62 miles) and it is at least 10 times better than the largest telescopes on the ground can achieve. Because of its fantastic ability to record fine detail, the Space Telescope will also be able to see much dimmer objects than can currently be detected. And since intensity of light falls away with distance, this means that astronomers will be able to see much deeper into the universe, observing galaxies that are seven times further away than anyone has seen before.

This single improvement in capability will effectively increase the volume of space that is visible to astronomers by a factor of 350. The Space Telescope will be able to detect objects as faint as the 28th magnitude (magnitude numbers increase as objects become fainter: the brightest star is about magnitude 1, the unaided human eye can see stars as dim as magnitude 6, and the best ground-based observatories can see to magnitude 23 or 24). The 28th magnitude corresponds to a distance of

The Space Telescope's primary mirror will focus starlight from long-dead galaxies close to the theoretical limit of observation.

about 14 billion light years: very close to the theoretical maximum distance that we can see from the solar system. And, because of the time light takes to travel, looking further out into the universe also means looking further back into the past. In a sense, the Space Telescope will be a time machine, revealing the universe in different stages of its evolution right back almost to the beginning of existence and the Big Bang itself.

As well as seeing deeper and more clearly into space, ST will offer another major benefit to astronomers. It will observe the universe not only in the part of the spectrum that the human eye can see, but also well into the ultraviolet and deep into the infrared. In fact, the Space Telescope will be able to focus light waves ranging in length from as little as 115 nanometers (in the far-ultraviolet) all the way up to as much as 1 million nanometers (or 1 millimetre, which is in the far-infrared). This enormous range is 3000 times wider than the visible portion of the spectrum, which only extends from 300 to 1000 nanometers. Some of the most interesting astronomical observations are to be made at ultraviolet wavelengths which are completely blocked by the atmosphere and hence impossible to see on Earth. It is here that the Space Telescope is likely to make some of its earliest discoveries.

## An observatory in the sky

The development that above all others has made the Space Telescope a reality is the Shuttle which will be used to launch ST and place it into orbit 500 km (310 miles) above the Earth. Throughout the telescope's planned 15-year lifetime, the Shuttle will pay regular return visits: servicing and maintaining its systems and repairing any components that may break down. In the event of a major failure, the Shuttle will even be able to take ST back on board and return it safely to Earth for a thorough refurbishment. The Shuttle is a sort of insurance policy safeguarding the future operation of the Space Telescope and it has allowed NASA's engineers to be ambitious in designing the spacecraft.

Of course, there will be no astronomers or astronauts actually on board the Space Telescope. Apart from the fact that images and other data from the spacecraft can be relayed to the ground perfectly well by radio, the presence of humans would greatly reduce ST's performance. Manned vehicles inevitably pollute space with waste gases which envelop the spacecraft like a small cloud (thus reintroducing the very atmospheric filtering effect that ST has been built to escape) and astronauts also ruin spacecraft stability by moving around – ST must point at stars for long periods with unprecedented precision.

Even without any living quarters, the Space Telescope is a large structure, taking up the full width and most of the length of the Shuttle's cargo bay. The spacecraft is 13.1 m (43 ft) long and 4.3 m (14 ft) wide, which is about twice the size of a railway locomotive. Once it has been deployed by the Shuttle, two large solar panels will open making ST seem larger still. Starlight enters the spacecraft through a door at one end of its cylinder-shaped body. This door can be closed to protect the incredibly sensitive instruments inside the telescope from receiving too much light. In fact, because of this danger, ST cannot be pointed within 50 degrees of the Sun, 15 degrees of the Moon or 70 degrees of the sunlit Earth. Although this restriction will limit the range of observations that ST can make at any particular moment, on a year-round basis all parts of

ST will be delivered into orbit by the Shuttle which will visit the observatory regularly throughout the telescope's 15 year operational lifetime.

Although smaller than many ground observatories, Space Telescope will see the Universe with unparalled resolution.

RIGHT ST during final assembly at Lockheed's Sunnyvale plant in California. For a sense of scale, note the technicians working on the spacecraft.

the universe will be open to its gaze. The only exception is the planet Mercury, which always lies too close to the Sun. Even in this case, however, it may be possible to use the Earth's shadow as protection, allowing brief observations of that hostile world.

At the heart of the Space Telescope are two mirrors that capture incoming light and focus it on to the scientific instruments. The primary mirror is 2.4 m (7.9 ft) in diameter and convex. It concentrates incoming light and reflects it 4.6 m (15 ft) back along the length of the telescope to a smaller concave secondary mirror 0.3 m (0.98 ft) in diameter. From the secondary mirror light bounces down the telescope again, passing through a small hole in the centre of the primary mirror. The beam then travels a further 1.5 m (4.9 ft) on past the primary mirror until it comes to focus on a collection of scientific instruments in the rear of the telescope that actually make the observations.

The mirrors themselves are as nearly perfect as current technology can make them. The optical surfaces are so flat that if the 2.4 m (7.9 ft) primary mirror were the size of Australia, then the tallest mountain or deepest valley would be knee-high to an ant! With this sort of precision special precautions must be taken to ensure that the mirrors retain their perfection. Once in space their temperature must be maintained within extremely precise limits and allowances must be made for the change in shape that occurs when the mirrors are no longer deformed by Earth's gravity. It will even be possible for ground controllers to calibrate the primary mirror in orbit by activating an army of needle-like bearings that can gently bend the mirror to correct its contour.

**Scientific instruments**

| *Instrument (by primary modes)* | *Field-of-view (arc seconds)* | *Resolution (arc seconds)* | *Band pass (nanometers)* | *Intensity range (stellar magnitudes)* |
|---|---|---|---|---|
| Wide field camera (f/12.8) | 2.17×2.17 | 0.1 | 115 to 1100 | 9.5 to 28.0 |
| Planetary camera (f/30) | 1.2×1.2 | 0.04 | 115 to 1100 | 8.5 to 28.0 |
| Faint object camera | | | | |
| (f/96) | 11.0×11.0 | 0.02 | 120 to 600 | 21.0 to 28.0 |
| (f/48) | 22.0×22.0 | 0.04 | 120 to 600 | 21.0 to 28.0 |
| Faint object | 0.1 to 4.3 | 3.0 Å | 115 to 700 | 19.0 to 22.0 |
| spectrograph | 0.1 to 4.3 | 30.0 Å | 115 to 700 | 22.0 to 26.0 |
| High-resolution | 0.25 to 2.0 | 0.03Å | 110 to 320 | down to 11.0 |
| spectrograph | 0.25 to 2.0 | 0.15Å | 110 to 320 | down to 14.0 |
| | 0.25 to 2.0 | 1.5Å | 110 to 170 | down to 17.0 |
| High-speed photometer | 0.4, 1.0, 10.0 | 16.0 μsec | 120 to 800 | down to 24.0 |
| Fine guidance sensors | 69.0×69.0 | 0.003 | 467 to 700 | 4.0 to 20.0 |

There are five scientific instruments on board Space Telescope, plus the Fine Guidance Sensors that can be used to determine the position of stars with great accuracy.

**Artificial eyes**

There are five scientific instruments on board Space Telescope, plus a system to point the spacecraft with great accuracy by locating and fixing on to known guide stars. These fine guidance sensors can also be used to make observations in their own right and so they really count as a sixth scientific instrument. The most important device is the wide field/planetary camera, which is the telescope's primary means of sending pictures down to Earth. As its name suggests, the instrument can operate in two different modes: a 'wide angle' configuration with a field of view of 2.67 arc minutes – an eye would fill the frame at a distance of 40 m (130 ft) – and a higher resolution planetary mode which yields images that are about three times more detailed, but with a correspondingly smaller field of view. In the latter case the camera will produce pictures with a resolution of 0.043 arc seconds within an image that is wide enough to include the full disc of a planet such as Saturn. Such pictures will be comparable in clarity to those taken by fly-by spacecraft such as Voyager just a few days away from closest encounter and they will be available on a regular basis year in, year out.

The pictures taken by the wide field/planetary camera will be recorded electronically on a light-sensitive array of detectors consisting of more than 2.5 million separate elements called pixels. This array is made up of four sophisticated silicon chips known as CCDs (charge-coupled devices) with a square arrangement of 800×800 pixels. Unlike conventional photographs, which require developing with chemicals, the pictures from this electronic camera will be available instantly. The CCDs turn the starlight focused by the telescope into signals encoding the intensity of the image at every pixel. The system can even take colour pictures by inserting filters into the path of the incoming light to record the different intensities at various wavelengths. This information subsequently allows a computer to reconstruct a coloured image.

Before such processing can take place, however, the data encoding the pictures must reach the astronomers waiting back on Earth. The signals from ST will take a circuitous route, first travelling up to one of NASA's Tracking and Data Relay Satellites in geosynchronous orbit before being re-transmitted to the White Sands receiving station in New Mexico in the United States. From here the signals will again be sent up to space, to be relayed by a commercial communications satellite to the Goddard Spaceflight Center in Maryland. Finally the data will travel the last few dozen kilometres by landline on to the Space Telescope Science Institute in Baltimore where it will be decoded, processed and archived.

The Space Telescope has another means of taking pictures to complement the wide field/planetary camera. The faint object camera (which has been built by the European Space Agency who are also supplying the solar panels) will produce images which exploit the maximum resolving and light-gathering power of ST's optics. It will detect faint

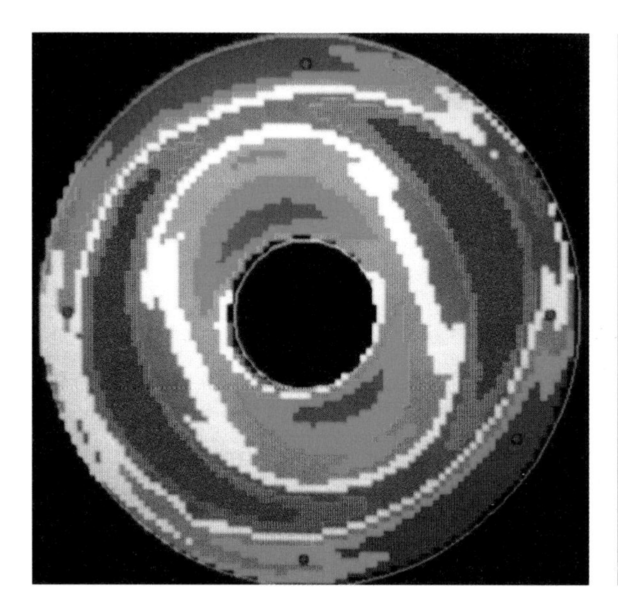

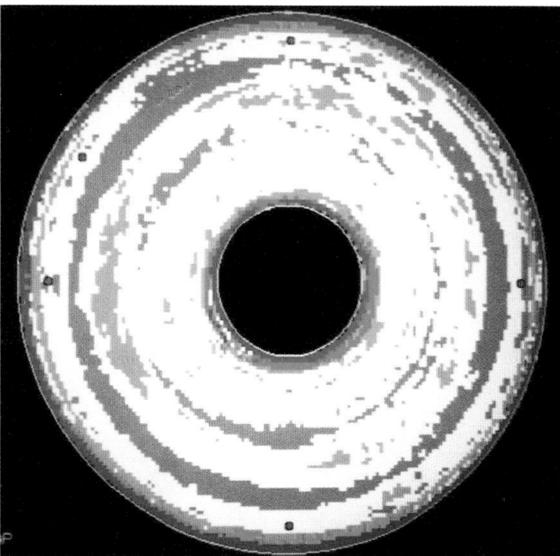

Before and after grinding: the colours represent small variations in the flatness of Space Telescope's primary mirror. After weeks of computer controlled polishing, the mirror's surface is almost perfectly smooth.

sources that no telescope has ever shown before and it will reveal an unprecedented level of detail in familiar astronomical objects. Instead of CCDs, the faint object camera employs a more conventional vidicon television tube and the incoming light is first amplified using an image intensifier. The instrument will be capable of registering individual photons of starlight and some exposures may take as long as 10 hours to produce a single picture of the faintest objects.

**Staying on target**

The requirements to take such high-resolution pictures places fantastic demands on the pointing accuracy of the Space Telescope. The spacecraft must be able to lock itself on to a chosen target and remain stationary with an accuracy of not less than 0.007 of an arc second. If a darts player could throw with such precision he would be capable of hitting a 1 cm bullseye from a distance of 300 km (185 miles)! And ST must remain locked on target for exposure times that can be measured in hours. Achieving this degree of control has pushed the technology of spacecraft guidance beyond all previous limits and has caused more problems than any other aspect of Space Telescope's development.

Normally spacecraft make use of gyroscopes to sense their motion in three-dimensional space and small thrusters to keep themselves pointed in the right direction. Unfortunately neither of these techniques is capable of achieving the sort of accuracy ST requires. Gyros are not sufficiently sensitive to detect very slight slow movements and thruster rockets would pollute the space environment – quite apart from the fact that the spacecraft would need to be replenished with fuel periodically. So instead ST uses two entirely different methods of sensing its position and maintaining its stability.

Not all of the light gathered by the primary mirror is directed to the five main scientific instruments. A substantial portion of the focused image is picked up by ST's three fine guidance sensors which use the information to keep the telescope stationary. To accomplish this feat, the computers forming part of the fine guidance system must know the expected position of previously selected guide stars which have been catalogued for every part of the sky. The fine guidance sensors tell the computer the actual position of the guide stars and by comparing this information with the theoretical position, the system can instruct the spacecraft to move until the two sets of data coincide. The necessary three-dimensional movement is achieved with reaction wheels (electrically driven devices rather like flywheels) that can be spun up or slowed down to reduce a corresponding rotation of the weightless spacecraft in the opposite direction.

The technique sounds simple enough in theory, but the practical difficulties involved in implementing it have been formidable. To indicate the sort of engineering precision required, the locomotive-sized telescope can be thrown off target merely by the reaction forces that are set up when one of its miniature data tape recorders stops or starts! In order to be sure of finding a pair of suitable guide stars for every possible target that ST may be required to observe, the project's astronomers have had to compile the most comprehensive computerized star map ever. One of the principal tasks of the Space Telescope Science Institute in the years before ST is launched has been to carry out this task.

The project's managers now admit that at the beginning of development they severely underestimated the difficulty of meeting the Space Telescope's stringent pointing specifications. As a result, early qualification tests of the Fine Guidance system were a failure. Launch of the telescope has slipped many years and costs have multiplied to well beyond the billion dollar mark. Several senior figures in NASA and at

Two views of Saturn: the photograph on the left was taken through one of the best ground-based telescopes; the right hand image shows a Voyager picture at well below anticipated ST resolution.

## The Range of the Space Telescope

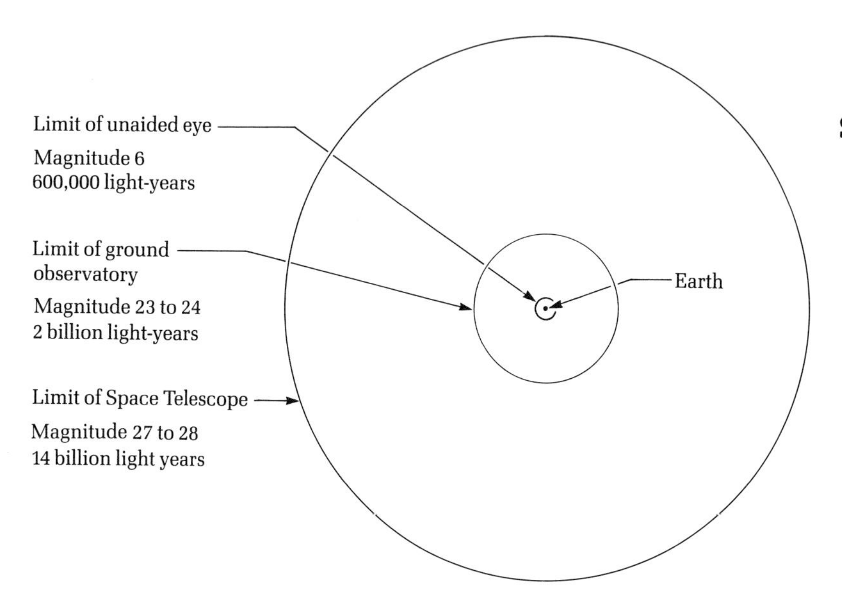


When ST becomes operational, the size of the universe open to optical astronomy will become 340 times larger.

the contractor companies were fired or have resigned. Following this low point in the telescope's history, vigorous efforts to remedy ST's ills have been made and there is now every indication that the stringent pointing requirement will be fulfilled.

**More than meets the eye**

The scientific instruments carried on board Space Telescope will be able to do far more than simply take pretty pictures. Two of the devices are spectrographs which can analyse light from a star or galaxy to discover an extraordinary amount about conditions far across the universe. They do this by separating starlight into its constituent wavelengths: rather as a prism can be used to resolve white sunlight into a rainbow spectrum. The exact mixture and balance of wavelengths in the light from a distant galaxy can reveal a wealth of information about the temperature, chemical composition, distance, and even the motion of the object in question. Although the results from the spectrographs may not look so spectacular as the pictures from the cameras, the information they provide will have enormous scientific significance.

The instrument with the greatest sensitivity, although with a lower resolving power, is the faint object spectrograph. It can respond to wavelengths from 800 down to 115 nanometers (well into the ultraviolet) and it is sensitive enough to yield usable spectrographs for stars as dim as the 18th magnitude, which is at least 10 stellar magnitudes fainter than any previous spectrographic observations at short wavelengths. For studies where the emphasis is on great precision, ST is equipped with the high-resolution spectrograph. At the expense of low-light sensitivity (and with a smaller band width from 320 to 110 nanometers), this instrument will be able to provide spectographs of incredibly precise regions of the sky. It will be able to detect chemical differences across the surface of a planet, distinguish between the separate stars in a binary pair, and analyse individual stars in crowded globular clusters – perhaps the most beautiful objects in the sky.

**Instrument astronomy**

| | |
|---|---|
| Wide field/planetary camera | General imaging, cosmic distance scales, cosmic evolution, the comparison of near and far galaxies, stellar population studies, the distribution of energy in stars and compact objects, the observation of stars in formation and supernovae, planetary atmosphere observations and comparisons, search for planets around nearby stars, cometary observations. |
| Faint object camera | Observe extragalactic supergiant stars, study variable brightness stars, gather data on globular clusters, examine binary star systems, search for extra-solar planets, establish stellar masses, do detailed studies of shock fronts and condensing gas clouds, search for direct evidence that quasars may be at the centre of faint galaxies. |
| Faint object spectrograph | Observe the nuclei of active galaxies, define the amount and kinds of chemicals in galaxies, provide information on the physical properties of quasars, study the mysterious jets that appear optically in photographs of quasars, study quasars believed to be at the nuclei of some galaxies, study comets before they are changed chemically by the Sun. |
| High-resolution spectrograph | Investigate the physical make-up of exploding galaxies, quasars and other dense objects, study the loss of mass of one star to another in binary systems, measure the total amount of matter expelled in stellar explosions, investigate the composition of interstellar and intergalactic gas clouds, study the various stages of stellar evolution, define the atmospheric structure of solar system planets, measure the chemical elements and investigate the structure of comets. |
| High-speed photometer | Make precise observations of rapidly pulsing compact objects, exploding variable stars and binary systems, examine properties of zodiacal light (sunlight reflected from the solar system dust) and diffuse galactic light, calibrate faint stellar objects and standard candles, examine specific spikes or flickers of light ejected from stellar objects including compact stars and supernovae. |

Some of the astronomical observations that will be made using ST's various scientific instruments.

Among the most intriguing observations that are possible with the Space Telescope's spectrographs will be the study of QSOs or quasars – mysterious objects that are thought to be the oldest visible things in the universe. The faint object spectrograph should be able to study very distant quasars to discover how fast they are travelling and what mixture of gases they contain. Both these items of information are crucial to cosmologists' understanding of the universe. Different models of the Big Bang predict that different ratios of hydrogen, helium and deuterium (an isotope of hydrogen) would have been present in the early universe. ST's superior spatial resolution and faint magnitude threshold will give it an enormous 'look-back' time. It will be able to make direct observations of the state of matter during the earliest epochs and provide the data to decide between rival theoretical accounts of how the universe began.

Paradoxically, the ancient quasars may also hold the key to our knowledge of the distant future. Quasars are receding from us in the

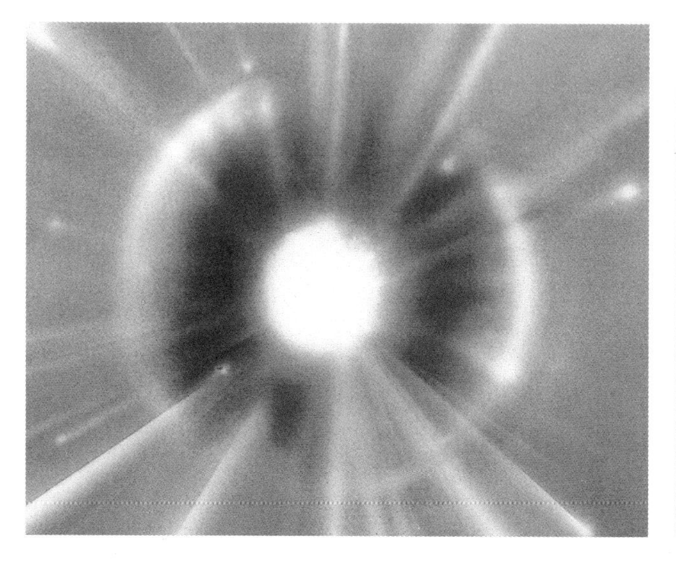

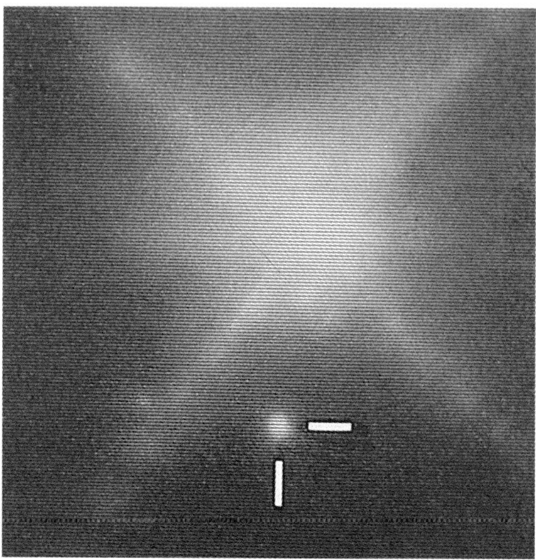

An artist's impression of a supernova (left). Scheduled observations will be cancelled at short notice to study one of these rare and spectacular stellar cataclysms. The right hand frame illustrates one of the most technically demanding early observations: the study of nearby stars to see whether faint planets can be detected close to the intensely bright parent sun.

aftermath of the Big Bang and the faster they are receding, the faster the universe as a whole must be expanding. The rate of expansion is known as the Hubble constant after Edwin Hubble, the American astronomer whose efforts in photographing hundreds of distant galaxies earlier this century established the fact that we live in an expanding universe. The Space Telescope, which is named after Hubble in honour of his discovery, will be able to measure the expansion with greater precision. If the Hubble constant is found to be above a critical value, then the universe will continue flying apart for ever, until all the stars go out and cold ashes are all that remain. If it is below, then gravitational attraction may be strong enough to halt the expansion, leading ultimately to the 'Big Crunch' and, some theoreticians believe, a cosmic rebirth. The Space Telescope may literally decide the fate of the universe!

The fifth scientific instrument on board ST is the High-Speed Photometer. It is designed to observe the brightness of a star or galaxy and measure any fluctuations in intensity with extraordinary accuracy. Changes on a timescale as brief as 10 millionths of a second can be distinguished – which is impossible with ground-based observatories because such minute changes are swamped by the atmospheric effects that make stars appear to twinkle. With no moving parts the High Speed Photometer is the simplest of the scientific instruments on board ST. But despite its simplicity, the device has the power to detect the smallest objects that can be resolved by the telescope. In principle, it will be able to observe a star only 3 km (2 miles) across: which is very near the threshold at which collapse to a black hole occurs. Such enormously dense objects can be expected to spin extremely fast and the High Speed Photometer will be able to measure the rapidly fluctuating output from these cosmic lighthouses. In addition, by accurately measuring the intensity of particular types of stars, the instrument can be used to calibrate the distance scale that astronomers have developed to map the universe.

The 'sixth' scientific instrument on board the Space Telescope is the Fine Guidance system which can pinpoint the location of stars to an

OPPOSITE Anatomy of Space Telescope: the observatory's main components as they will be arranged in the orbiting spacecraft.

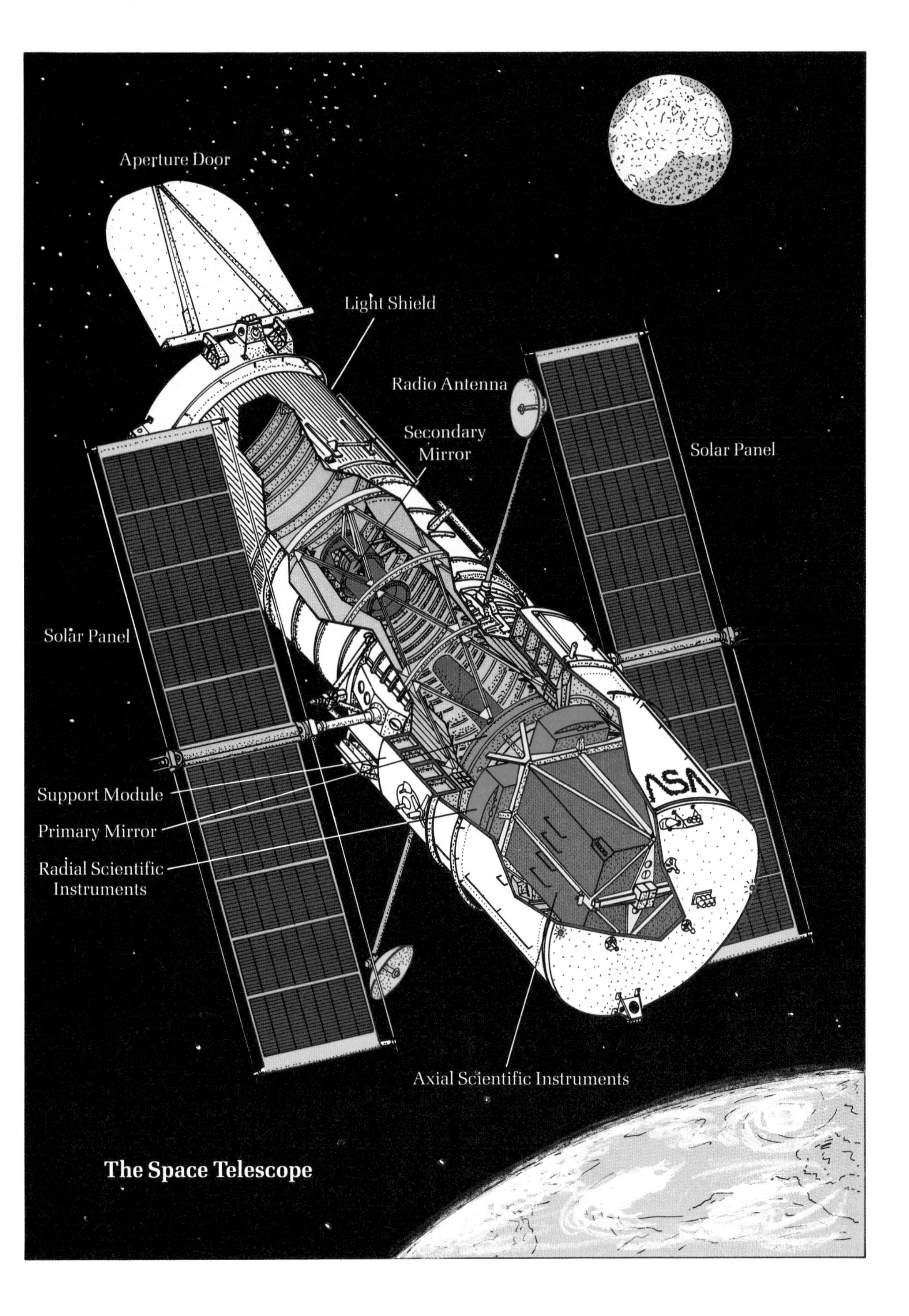


**The Space Telescope**

accuracy of 0.002 arc seconds. Although their primary function is to keep ST locked on target, so that the other scientific instruments can carry out their measurements, the fine guidance sensors can also make valuable observations in their own right. In particular, using a technique called astrometry, they will be able to measure stellar distances 10 times more precisely than is now possible. And the Fine Guidance system may provide an answer to perhaps the most exciting question of all: are there planets around stars other than our own Sun?

It is just possible that the faint object camera, at the limit of its spatial resolution, may be able to spot Jupiter-sized planets directly. This is, however, a dangerous task because the bright light from the nearby star could actually destroy the detectors. But there is another approach: the Fine Guidance Sensors should be able to detect the 'wobble' that occurs as a star's course is perturbed by the orbit of a large companion planet. This method relies on comparing the exact position of a star over a period of months or even years, and there are plans to make regular observations of 500 or so nearby stars to determine how many of them have planetary systems. The answer could be a high proportion, or almost none. In either case the implications are astonishing. If planetary systems are frequent, then we can assume that the conditions for life to develop are relatively common and so there are probably a large number of sentient races in the universe. If planetary systems are rare, then life becomes all the more precious, and we must consider the possibility that we may be alone.

**Scheduling a revolution**

The Space Telescope offers such a dramatic improvement over existing observatories that its services will be in demand by virtually every optical astronomer in the world. ST will be making observations round the clock for at least 15 years, but even so there will be far more demands on its time than there are hours available. The over-subscription rate is expected to exceed 15:1. The difficult task of allocating time on the telescope and scheduling all the different observations will be handled by the Space Telescope Science Institute at Baltimore in the United States. Astronomers from all over the world will visit the Institute for several weeks at a time to work with ST and make a preliminary analysis of their results. They will leave with a magnetic tape under their arms containing data which remains secret to them for a period of one year, allowing time for relatively unhurried publication in the cut-throat world of astronomy! The teams that developed the scientific instruments have a special privilege. They have each been awarded 360 hours of precious observation time on ST, which virtually guarantees them the honour of making some major astronomical discoveries.

Schedules will be worked out at least six months in advance to make most efficient use of the telescope. It takes significant time for the spacecraft to move from one target to another (18 minutes for 90 degrees, which is about the same rate as a minute hand sweeps round the face of a clock) and the pointing system also requires several minutes to lock on to new guide stars. Planners must group together different observations in the same part of the sky. Gaps will be left in the schedule for 'targets of opportunity': perhaps a sudden supernova in another galaxy, a newly discovered comet, or some aspect of an earlier ST observation that demands further study. Riccardo Giacconi, Director of the Space Telescope Science Institute, has up to 15 per cent of ST's time available at his discretion for allocation as he thinks best. In astronomical circles he is a powerful man!

During the planned 15-year life of the Space Telescope the Shuttle is likely to pay numerous servicing visits. The solar panels are expected to need replacing approximately every five years and a number of other spacecraft components – particularly the batteries, gyroscopes and reaction wheels – may be expected to fail as time goes by. Almost all the major systems have been designed for on-orbit replacement by astronauts; NASA is increasingly moving away from the idea of a planned return to Earth for refurbishment. Such a move remains technically possible if problems develop with the telescope's mirror or radio transmitters, but it would be expensive and it would mean taking ST out of action for well over a year.

Atmospheric drag is another potential problem facing ST, particularly during periods of high solar activity when the Earth's outer atmosphere becomes more dense. The spacecraft will initially be placed in the highest possible orbit, probably around 520 km (320 miles) depending on the performance of the Shuttle's main engines. ST will inevitably lose height as it collides with molecules of gas, but its operations will be unaffected until its orbit decays below 480 km (300 miles). At this altitude, minute aerodynamic forces on the solar arrays will begin to disturb pointing accuracy, and a Shuttle-assisted reboost will be necessary. During the peak of the sunspot cycle, such visits may be required as often as every nine months.

As well as ensuring reliable service, the Shuttle will also allow the scientific instruments on board ST to be changed and updated. All five instruments are contained in standard modules that can be removed by astronauts to be exchanged with replacements. This capability is all the more impressive given that each removable module, which is about the size of a telephone booth, must be optically aligned with fantastic precision. NASA has already made an advance announcement of opportunity for teams to propose a second generation of devices. It is likely that one of the earliest changes will be to install an infrared camera, which is an area where the coverage provided by the current complement of scientific instruments is weakest. Furthermore, new state-of-the-art devices can be installed to keep pace with technological improvements. ST will remain at the leading edge of astronomy for years into the future.

ST is designed to be serviced by visiting astronauts in orbit. All the scientific instruments, as well as many critical spacecraft sub-systems, can be easily removed and replaced.

It is hard to predict exactly where Space Telescope will have its greatest impact. The spacecraft has been described as the modern equivalent of the magnificent cathedrals that were built in Europe during the Middle Ages. ST represents the very best that 20th-century technology can achieve and its reward will be a wealth of new information, discovery and enlightenment. We can anticipate many of the scientific insights that the new capability will bring, but the most exciting discoveries will be the ones we cannot predict. The Space Telescope will open wide our window on the universe and no one knows what strange truths it may reveal.

CHAPTER SIX

# Industry in Space

*Once an exotic idea in the realm of science fiction, space as an arena for a broad spectrum of commercial activity is coming into reality.*
Editorial. *Aviation Week & Space Technology*

Space, it seems, is no different from any other frontier: the story is the same whether oceans, mountains or high vacuum are the obstacles for mankind to overcome. First, pioneers explore the new territory to map the land and discover its resources. Then a transportation system is established so that people and goods can journey there with ease. Finally, the era of exploitation begins, as industry moves in and attempts to take advantage of the new-found riches. Space operations are now entering this third commercial phase and during the next ten years there will be a massive expansion of industrial activity in orbit. Big business is moving into space.

According to the Center for Space Policy, an American organization that advises companies interested in commercial opportunities in space, the turnover from space industries in the year 2000 will be a massive $65 billion. More conservative estimates put the total at around

The view from space: ripe for commercial exploitation.

$40 billion. But even the lower forecast implies the creation of a major new sector of the world's economy in little over a decade – so where is all this money going to come from? The question becomes all the more puzzling given that, by definition, the new industrial zone will be located in empty space.

Although there are certainly no material resources in Earth orbit (where industrial operations will initially be confined), there are less tangible assets which can be exploited. In particular, an orbiting spacecraft has two unique and valuable characteristics. First, a satellite flying high above the Earth commands an exceptional view of the ground. This observational advantage has applications in communications, remote sensing, and navigation. Second, because an orbiting spacecraft is in a state of free fall, everything on board is weightless. Zero gravity enables a range of potentially lucrative manufacturing processes that would be impossible back on Earth.

On these foundations an entire new industry is now being built. At the latest count more than 50 companies, most of them recently formed, are drawing up plans to make money in space. These industrial pioneers sense an enormous commercial opportunity, not only for those who operate the satellites and the factories, but also for a whole range of new companies offering space services that have traditionally been supplied by government agencies like NASA: services that will include everything from launch to on-orbit maintenance.

**The following is a selection of companies involved in commercial space-related operations.**

|  |  |
|---|---|
| Aluminum Co. of America | Interested in producing unique metal alloys, such as aluminium lithium, that will not mix properly on Earth. |
| Astrotech International | Various major launch service operations ranging from Shuttle payload processing to the proposed purchase of a complete fifth Orbiter. Also working with McDonnell Douglas on the development of a new Shuttle upper stage. |
| Commercial Cargo Spacelines | Comprehensive Shuttle launch services from insurance to payload integration. |
| Eastman Kodak | One of the seven companies that have bid to take over the US Landsat remote-sensing spacecraft. |
| Fairchild Industries | Developing an unmanned multi-orbiting platform called Lasercraft to be launched on the Shuttle. Fairchild will rent out room on the platform to other companies for a variety of commercial operations. |
| General Dynamics Convair | Hoping to begin commercial production of the highly successful Atlas Centaur expendable booster. |
| Geostar | Intending to launch a series of satellites for low-cost personal communications and navigation. |
| Getaway Special Services | Provides design and installation services for universities and other groups using the Shuttle's Getaway Special canisters. |

| | |
|---|---|
| John Deere | Undertaking metallurgical experiments on board the Shuttle aimed at improving ground-based production techniques |
| Johnson & Johnson | Carrying out commercial production of pharmaceuticals using the EOS system developed jointly with McDonnell Douglas. |
| Lovelace Medical Foundation | Planning Shuttle-based experiments to improve production and purification of monoclonal antibodies for diagnosis and treatment of disease. |
| McDonnell Douglas Astronautics | Developing hardware and support systems for EOS (Electrophoresis Operations in Space) drug production and also marketing the PAM Shuttle upper stage booster. |
| Microgravity Research Associates | Hoping to manufacture highly pure gallium arsenide crystals in orbit for the electronics industry. |
| 3M (Minnesota Mining & Manufacture) | Planning experiments on the Shuttle aimed at making organic crystals and thin films in space. |
| Orbital Sciences | Designing various Shuttle upper stages for commercial orbit transfer applications. |
| Space Industries | Proposing to develop a free-flying industrial platform which would be launched and serviced by the Shuttle. |
| Space Services of America | Developing the Conestoga expendable booster for small payloads. |
| Sparx | Intending to offer commercial remote-sensing services based on the SPAS Shuttle pallet. |
| Spot Image | European operation that will use the French Spot satellites to provide 10 m resolution stereoscopic ground imaging. |
| Transpace Carriers | Undertaking commercial marketing of the highly successful McDonnell Douglas Delta launch vehicle. |
| Wyle Laboratories | Considering development of a manned orbiting laboratory module which would be attached to the Space Station and would be rented out to industrial users. |

Arthur C. Clarke identified the telecommunications potential of geostationary satellites in a 1945 article in the British magazine 'Wireless World'.

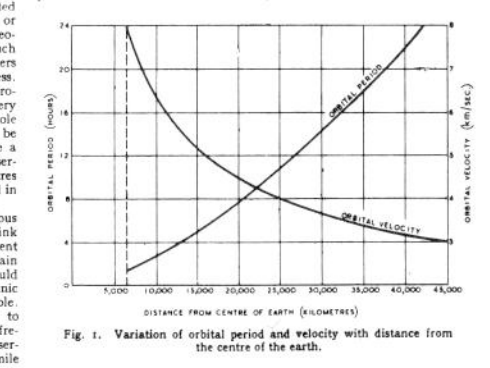

**Wireless World** 305

XTRA-TERRESTRIAL RELAYS

*ocket Stations Give World-wide Radio Coverage?*

*By ARTHUR C. CLARKE*

t is possible, by
choice of fre-
l routes, to pro-
circuits between
r regions of the
part of the time,
mmunication is
by the peculiar-
phere, and there
ns when it may
A true broadcast
constant field
times over the
ld be invaluable,
spensable, in a

though the tele-
raph position is,
n is far worse,
ic transmission
yed at all. The
elevision station,
good site, is only
miles across. To
country such as
ald require a net-
itters, connected
, waveguides or
. A recent theo-
shown that such
require repeaters
fty miles or less.
kind could pro-
verage, at a very
, over the whole
ry. It would be
ion to provide a
with such a ser-
the main centres
ald be included in

s equally serious
t is made to link
ces in different
e. A relay chain
miles long would
and transoceanic
till be impossible.
rations apply to
f wide-band fre-
ion and other ser-
gh-speed facsimile
their nature re-
e ultra-high-fre-

nsider the solution
discussion too far-
en very seriously.
is unreasonable,
nvisaged here is a

logical extension of developments in the last ten years—in particular the perfection of the long-range rocket of which V2 was the prototype. While this article was being written, it was announced that the Germans were considering a similar project, which they believed possible within fifty to a hundred years.

Before proceeding further, it is necessary to discuss briefly certain fundamental laws of rocket propulsion and "astronautics." A rocket which achieved a sufficiently great speed in flight outside the earh's atmosphere would never return. This "orbital" velocity is 8 km per sec. (5 miles per sec), and a rocket which attained it would become an artificial satellite, circling the world for ever with no expenditure of power—a second moon, in fact.

the atmosphere and left to broadcast scientific information back to the earth. A little later, manned rockets will be able to make similar flights with sufficient excess power to break the orbit and return to earth.

There are an infinite number of possible stable orbits, circular and elliptical, in which a rocket would remain if the initial conditions were correct. The velocity of 8 km/sec. applies only to the closest possible orbit, one just outside the atmosphere, and the period of revolution would be about 90 minutes. As the radius of the orbit increases the velocity decreases, since gravity is diminishing and less centrifugal force is needed to balance it. Fig. 1 shows this graphically. The moon, of course, is a particular case and would lie on the curves of Fig. 1 if they were produced. The proposed German space-stations



**Fig. 1. Variation of orbital period and velocity with distance from the centre of the earth.**

The German transatlantic rocket A10 would have reached more than half this velocity.

It will be possible in a few more years to build radio controlled rockets which can be steered into such orbits beyond the limits of

would have a period of about four and a half hours.

It will be observed that one orbit, with a radius of 42,000 km, has a period of exactly 24 hours. A body in such an orbit, if its plane coincided with that of the

## The view from space

The excellent vantage point offered by an orbiting satellite was appreciated well before Russia launched the first Sputnik in 1957. The British science fiction writer and engineer Arthur C. Clarke suggested a global communication system based on space relays back in 1945. He proposed placing satellites in a 24-hour equatorial orbit, so that they would appear to stay fixed in the sky as the Earth turns on its axis. Line of sight communications would be possible between any two points on the Earth simply by bouncing radio signals off the spacecraft. Much to his own misfortune, Clarke never patented the idea. Today there are over 300 such satellites in geosynchronous orbit and their inventor has missed becoming one of the richest men in the world.

Communications satellites are now a familiar and pervasive feature of modern life. Direct-dial international telephone calls and live television coverage of events across the world are so commonplace that they are taken for granted. The global satellite network also underlies a less visible information revolution that has transformed the operations of banks, multi-national corporations, news organizations and governments. Telecommunications satellites are now a billion-dollar-a-year business that has become the first (and so far the only) successful commercial space industry. In the coming years its scale and importance will grow with the proliferation of data links between computers, direct broadcast television and video conferencing.

The Center for Space Policy forecasts that by the year 2000 the space communications market will have grown to reach an annual turnover of $15 billion. To meet this demand the West will need to orbit 200 to 400 new telecommunications satellites in the civilian sector alone. With launches costing approximately $40 million per satellite, this represents an enormous commercial opportunity for companies able to deliver payloads to space. The satellite launching business promises to be one of the most competitive space industries in the coming years.

The market will be dominated by two principal contenders: NASA's Shuttle and the European Space Agency's Ariane. At current pricing levels, their charges for delivering a satellite to geostationary orbit are roughly equivalent. Both systems are heavily subsidized by government, and both are likely to become more expensive in the future as their prices are increased to reflect actual costs more closely. In the long run Ariane, being unmanned and expendable, is likely to win out over the Shuttle in purely financial terms. The Shuttle, on the other hand, will be able to offer a more sophisticated service, with options for on-orbit servicing by astronauts and return of payloads to Earth.

However, both NASA and ESA will be facing competition from elsewhere. The Japanese are developing a family of large boosters capable of delivering satellites to geostationary orbit. They are only waiting for the Americans and the Europeans to tire of subsidizing launches with taxpayers' money before competing on a real commercial basis. NASA will also be up against some home-grown rivals. In 1984 the Reagan administration passed legislation that makes it much easier for private companies in the United States to launch their own rockets. NASA is obliged to provide firing range facilities and there are some attractive new tax incentives for space businesses. Several American companies have been created to cash in, many of them with senior figures recently retired from NASA on the board of directors. One such company, the Space Shuttle of America Corporation, is even proposing to buy and run one or more of NASA's Shuttle Orbiters. Most of the bidders, however, are more interested in winning a share at the lower end of the market.

Transpace Carriers Incorporated, for example, has acquired exclusive marketing rights to the highly successful Delta booster, which can launch a payload of approximately 1400 kg (3085 lb) to geostationary orbit. This is substantially less than Ariane or the Shuttle, but there are nevertheless plenty of satellites in this class that will need launching between now and the end of the century. Space Services of America Incorporated, run by former astronaut Deke Slayton, is developing its own booster called the Conestoga. Initially this rocket will only be able to lift a small payload to low-altitude orbit, but the company has ambitious plans for larger versions.

The company also has plans to launch a rather unusual cargo using its new booster. Space Services of America has teamed up with the Celistis Group, a Florida-based consortium of morticians, to plan what will surely go down as the most tasteless mission in the history of spaceflight. For around $3000 the company will cremate a dead human body and place the ashes in a special capsule about 1 cm in diameter and 4 cm long. Several thousand of these capsules will then be sealed inside a highly reflective spacecraft and launched by the Conestoga booster for 'burial' in an orbit that should not decay for at least 63 million years. In the meantime (on clear nights) relatives and descendants of the deceased will be able to see their loved ones passing overhead like shooting stars.

Europe's Ariane booster (left) is the Shuttle's main rival in the Satellite launching business. Commercial contenders include the smaller and relatively unproven Conestoga Booster that may be employed in a macabre mission to 'bury' pelletised human remains in orbit!

There are less macabre opportunities for companies to provide commercial services out in orbit. The Fairchild Industries company is building an unmanned vehicle called Leasecraft which is launched by the Shuttle and collected after spending several months out in space. Fairchild will rent room on the Leasecraft to other companies who are interested in getting their payloads into orbit without having to develop a complete spacecraft. The first flight of a Leasecraft is expected to take place in 1987.

In a similar way, numerous ventures have sprung up to help other people use the Shuttle. If you have a satellite to launch then Commercial Cargo Spacelines will handle everything from payload integration to insurance. If it is a small experiment, then Getaway Special Services will fit it into one of the small canisters that can be booked on to the Shuttle at bargain basement rates. There is even one company, Wyle Laboratories, that is hoping to build its own modest Shuttle-launched space station, which universities and research organizations would hire for their own missions. Astrotech International is designing a space ferry that could carry satellites from one orbit to another.

Following on the success of the satellite telecommunications industry, space transportation looks set to be the next profitable operation in orbit. But there are plenty of other ideas in the pipeline, and communications satellites will not be the only type of vehicle looking for a ride into space. An orbital vantage point offers other commercial possibilities. Remote-sensing applications, including weather forecasting, analysis of land use, crop monitoring, geological prospecting, ocean surveying, pollution control, and even urban planning, are set to become commercial sector operations. The American government is selling off NASA's Landsat series of remote-sensing satellites and the Reagan administration has stated that future Earth resource studies must increasingly rely on private enterprise funding. At least half a dozen companies intend to develop this market, ranging from long-established multi-nationals like Eastman Kodak, to newly created specialist firms with futuristic names like the Earth Observing Satellite Company and Space America Incorporated.

Navigation is another area where industry is hoping for profits from space. A company called Geostar, based in Princeton, New Jersey, is planning to establish an ambitious satellite system for sending precise information about position to small and inexpensive portable terminals anywhere on the ground. The company hopes that the terminals will sell for as little as \$300 to \$400, which is at least ten times cheaper than existing (and far less portable) equipment. The system would not only inform users of their exact location (providing map references accurate to a few metres); it could also be used to exchange short messages via the orbiting satellites. Such a science fiction device would obviously be popular with people who need to find their way and keep in touch in remote or unfamiliar locations. The system could also be used in con-

TOP Getaway special canisters hitching a low-cost ride in the Shuttle's payload bay.

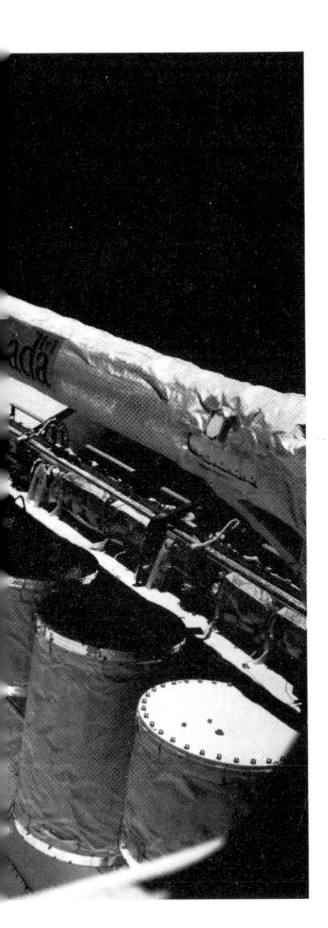

junction with small laser discs to provide a computerized road map for cars that would show the driver his exact position. Japanese companies are reported to be developing the necessary technology.

**Made in space**

All the commercial activities so far described are basically extensions of existing government-funded operations. Potentially the most significant and exciting space industry, however, is an entirely new kind of undertaking: the manufacture of high-value substances using technology that exploits zero gravity. According to the forecast by the Center for Space Policy mentioned earlier, by the year 2000 the revenue from this sector of the space economy will reach more than $40 billion. Products with the label 'made in space' will include life-saving drugs, computer chips, fibre optics, metal alloys and magnetic tape. All will have a very high value and all will be manufactured using techniques that would be impossible on the ground.

On Earth gravity rules the physical world. It does more than keep our feet planted firmly on the ground; gravity is responsible for a whole range of seemingly unrelated phenomena. Flotation, convection and sedimentation, the three curses of materials-processing technology, are all driven by gravity. These disruptive influences predominate at the microscopic level. They introduce irregularities into even mixtures and they overwhelm more subtle forces that could otherwise be gainfully employed. Remove gravity, and at a stroke you remove one of the main barriers to improved efficiency in numerous industrial processes that isolate and purify valuable substances.

BOTTOM A load of balls – the first made-in-space commercial product.

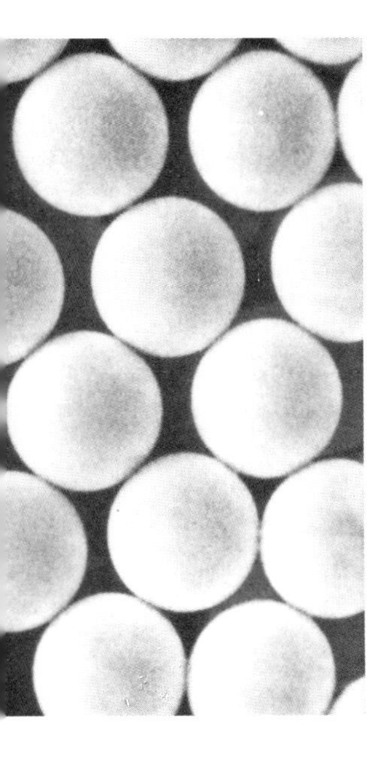

Out in space (contrary to widely held belief) gravity is *not* removed, instead its *effects* are no longer experienced. A spacecraft orbiting the Earth is still very much under the influence of our planet's gravity. Indeed, it is permanently falling towards the ground. But because the spacecraft is also moving forward, it maintains its altitude by following the curve of the planet. Since there is nothing to oppose the force of gravity, objects in orbit have no apparent weight (they do still have mass) and so people and loose items appear to 'float'.

The important thing from the industrial standpoint is that weightlessness eliminates virtually all convection, flotation and sedimentation. The effects of gravity are reduced 1000 times at least (astronauts' movements and attitude control thrusters cause minor disturbances, which is why the term microgravity is often preferred). Separation and isolation processes not only work more efficiently in space, but they also produce substances with a degree of purity that is impossible on Earth. Zero gravity techniques can also be used to manufacture completely new kinds of material: for example, alloys made from metals that will not mix in terrestrial foundries. Containerless manipulation of large samples, which can result in highly uniform structures, is also feasible in weightlessness.

The idea of factories in space is not just pie in the sky. There are already successful examples of commercial products manufactured under weightless conditions on the Shuttle. the first 'made in space' product ever to be sold was actually a load of balls: billions of microscopic polystyrene spheres each measuring *exactly* 10 millionths of a metre across. In fact their precisely determined size is the point. They are intended to provide a standard reference measure for use under microscopes, for calibration of filters and in developing high-precision optics. Batches of 15 g (0.5 oz) of the spheres were produced on several early Shuttle missions in an automated device called the monodisperse latex reactor. Free from distortions caused by Earth's gravity, the process produces enormous numbers of identical and perfectly round spheres. Samples have been sold to universities and other scientific organizations at a price that equates to $23 million a kilo. The space balls are worth nearly 2000 times their own weight in gold! Unfortunately NASA is unlikely to make much of a profit from this pioneering venture: demand for these specialized items is rather limited.

## Space medicines

The second successful manufacturing process to have flown on the Shuttle holds much greater promise. McDonnell Douglas and Johnson & Johnson have formed a partnership to produce very high value drugs in space under a programme known as EOS (Electrophoresis Operations in Space). Under the terms of a Joint Endeavour Agreement with NASA, EOS units of various descriptions have been flown on numerous Shuttle missions and McDonnell Douglas has even been allowed to send its own engineer, Charlie Walker, up on the Shuttle to operate the equipment. The EOS system is capable of purifying a large range of biological substances with a yield almost 1000 times greater than on the ground. The resulting product is also up to five times purer than can otherwise be achieved.

Electrophoresis works by passing an electric field across a mixture of substances to be separated. (Usually this mixture is a weak solution of the valuable target drug, swimming in a sea of unwanted, but chemically indistinguishable, impurities.) As the mixture flows through the field, the various substances are each displaced by an amount that is critically dependent on its molecular mass. The stream emerges from the electric field neatly banded into its different constituents, which can then be individually collected in a long line of containers. On Earth, the efficiency of electrophoresis is severely limited by gravitational effects (particularly convection), which constantly disturb the mixture and blur the different bands. In space, this problem is almost completely eliminated, resulting in the spectacular improvements noted above.

Because of the high costs of operating in space, EOS is best suited to the production of very high value drugs which are difficult to manufacture on Earth and which are effective in small doses. Proteins fit the bill

BOTTOM McDonnell Douglas engineer Charlie Walker has flown several times on the Shuttle to tend the EOS drug purification system (left of frame) that he helped to develop.

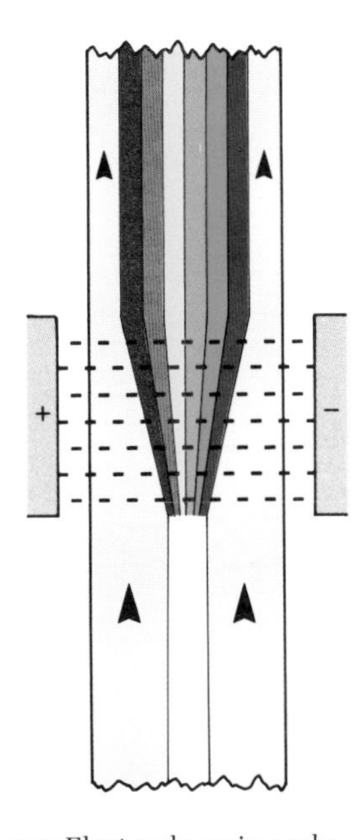


TOP Electrophoresis works by using an electric field to separate molecules of different sizes: smaller molecules are displaced more than large ones and purified bands of the different constitutents emerge.

admirably, with such substances as urokinase (an enzyme for treating heart disease), Factor 8 (the missing clotting agent in the blood of haemophiliacs), and interferon (an experimental anti-cancer drug) currently costing hundreds and even thousands of dollars a dose. There are also hopes that EOS might be used to isolate human pancreas beta cells, which could be the key to providing a permanent cure for some types of diabetes. However, McDonnell Douglas and Johnson & Johnson are being secretive about exactly which products they have been experimenting with. All they are prepared to reveal is that their hopes rest on a valuable human hormone that is almost impossible to make on Earth.

So far only prototype EOS production units have flown in space. McDonnell Douglas will shortly experiment with a larger 2000 kg (4400 lb) device situated in the Shuttle's payload bay, rather than in the crew mid-deck area where the system currently resides. The new automatic unit can operate continuously while the Shuttle remains in orbit, and although the device is still too small for commercial production, it will be able to purify sufficient quantities of a drug to carry out large-scale clinical trials. In the long run, McDonnell Douglas has plans to build a large-scale automated factory which would be left in orbit by the Shuttle to work unattended for months at a time. The company also intends to experiment with the EOS system on Leasecraft and even to build a dedicated Space Station module if the process is a real commercial success.

Unfortunately, however, the EOS project suffered a major setback in late 1984 when it was discovered that the entire batch of purified material returned from Shuttle Mission 41-D had been contaminated by a bacterial infection and rendered useless. This problem was probably due to the fact that launch had been delayed several times, giving bacteria additional opportunities to grow in supposedly sterile fluids. The failure of the system on this occasion does not indicate that there is anything wrong in principle with EOS, but it does indicate the high risk of undertaking complex industrial processes in orbit.

### Getting off the ground

Despite such risks, analysts are nevertheless confident about the ultimate potential for profit. The Center for Space Policy predicts that manufacture of pharmaceuticals will be the single largest space industry by the year 2000, with an annual revenue of $27 billion. Rockwell International, the huge American aerospace contractor, forecasts $20 billion. It seems that pills made in space are going to be big business. There are, too, other high-value substances that are attracting the attention of new space companies.

Microgravity Research Associates (MRA) was the second company to sign a joint endeavour agreement with NASA (McDonnell Douglas and Johnson & Johnson were the first). It intends to manufacture high purity gallium arsenide crystals for use in the next generation of super-fast computer chips. Under the agreement, MRA is entitled to eight free rides on the Shuttle, in return for which NASA will get the chance to demonstrate the commercial potential of its transportation system without paying for any research and development. NASA hopes that its partners will subsequently buy further time on the Shuttle at the standard rates. MRA is planning to use a low-temperature furnace to grow its gallium arsenide crystals. The technique should eliminate gravity-induced imperfections which seriously degrade the performance of large crystals grown on Earth, and the semiconductor industry is expected to pay a high price for the new product.

There is, however, a question mark hanging over the future of these and other proposals to manufacture new products in space. The enormous cost of operating factories in orbit means that the value of whatever is produced must be enormous: as much as a million dollars per kilogram according to some estimates. This severely restricts the range of products that it is currently worth considering for commercial production. Foamed metal could be made in space, for example, and although it would be incredibly light for its strength, it could not possibly be worth the manufacturing cost. Furthermore, ground-based alternatives are constantly being refined and improved. There is already a new electrophoresis process being developed at Harwell in Britain that threatens to compete directly with McDonnell Douglas and Johnson & Johnson's far more costly EOS system.

Cost is not the only consideration: the risk of operating in space is another factor likely to dampen the enthusiasm of many businessmen. One consequence of the Shuttle's recent problems launching satellites is that the spacecraft owners are finding that insurance premiums have gone through the roof. Following the loss of the US Navy's Leasat on mission 51-D in April 1985, insurers were quoting 20 per cent of the payload value as the minimum they would accept to cover future Shuttle launches. With a typical communications satellite like Leasat valued at $85 million, that is quite a price to pay!

Many people involved in the business are now warning that there is a danger of overselling commercial opportunities in space. This is particularly true with NASA loudly trumpeting the industrial potential of the Space Station. Companies are being enticed to test the water with the inducement of free rides on the Shuttle, and without such incentives it is doubtful that any of the present ventures would be taking place. Obviously NASA does have a legitimate role in getting the incipient enterprise off the ground, and the current surge of industrial interest in space is certainly worth sustaining with practical demonstrations of the technology. But there is a very long way to go before a profitable industry is established. The opportunity is undoubtedly there and the commercialization of space promises to be one of the most interesting developments during the coming decade.

## CHAPTER SEVEN
# War In Space

*It is one of the tragic ironies of our age that the rocket, which could have been the symbol of humanity's aspirations for the stars, has become one of the weapons threatening to destroy civilization.*

Arthur C. Clarke, writer and visionary

Although both superpowers are anxious to stress the peaceful aspects of their respective space programmes, there is no escaping the pervasive involvement of the military in space operations. Sputnik 1, the world's first artificial satellite, was launched on 4 October 1957 atop a Russian missile originally intended to carry a nuclear warhead. Similarly, America's Explorer 1, which went into orbit on 31 January 1958 was the result of a project run by the United States Army. Today, the armed forces of both nations maintain an extensive presence in space and expenditure on military applications outstrips spending on all civilian space activities combined. Modern warfare relies heavily on space technology and the indications are that space itself is set to become the battlefield of the future.

### The high ground

There are two straightforward reasons why space is strategically important. First, the military organizations of both superpowers, and indeed of all NATO and Warsaw Pact countries, are today highly dependent on orbiting satellites to supply a range of vital services. At the latest official count, the United States has 'more than 40 operational Department of Defense spacecraft on orbit performing missions'. These satellites provide numerous facilities including early warning of enemy missile

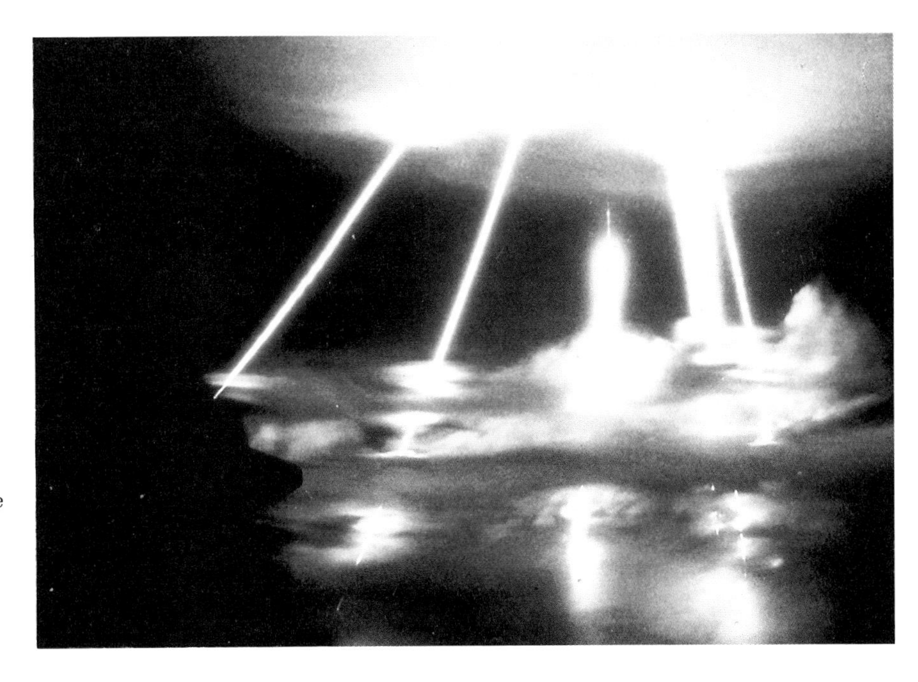

A vision of Armageddon: multiple nuclear warheads from an incoming MX intercontinental ballistic missile re-enter the atmosphere in a test high above the Pacific Ocean. America's Strategic Defense Initiative requires the development of directed energy weapons that could destroy these fast-moving targets before they reach their destination.

launches, radar surveillance of ships at sea, electronic eavesdropping, photo-reconnaissance of foreign territory, navigation for surface troops, aircraft and missiles, provision of weather data and, above all, long-distance communications. It is no exaggeration to say that without these satellites a modern fighting force would be crippled.

The second reason why the military have such a strong interest in space stems from the perception that, as the necessary technology becomes available, it will become feasible to install powerful and sophisticated weapon systems in orbit. From the vantage point of space, such weapons could ensure strategic domination of all military activity on the surface and in the air. Such a dream – or nightmare if you prefer – will require the development of new kinds of weapon, such as directed-energy beams and electromagnetic cannon. It will also require fantastically effective surveillance systems to monitor myriad events below, and incredibly powerful computers to make battlefield decisions that would be far beyond the capacity of any human being. This vision is not merely science fiction fantasy. As discussed later in this chapter, the United States is spending tens of billions of dollars in an attempt to convert just such a scenario into reality.

Soldiers in space: the new high ground.

The Pentagon's attitude to space was bluntly stated in a recent article by Edward Aldridge, Under Secretary of the United States Air Force. Writing in the Pentagon's *Defense* magazine he said:

> For many, space is the 'ultimate high ground', the last and greatest frontier that lies before man. For the military, if we accept this premise, space is a 'fourth medium' to complement military actions on land, sea and in the air. We do not have to stretch our imagination very far to see that the nation that controls space may control the world. We must be sure that the Soviet Union is not proceeding down that path, or if they are, we must ensure that they do not achieve that dominance.

As if in reply, a statement was broadcast on Radio Moscow a short while later: 'It would be naïve to believe that a powerful nation such as the Soviet Union, the pioneer in space exploration, would allow the United States to achieve military supremacy in space.' Thus the battle lines are drawn. Each superpower is militarizing space to prevent the other from doing so first, and earthly conflicts spill over into the new arena.

## Soldiers in space

The United States has not been slow to establish a secure military footing in space. On 1 September 1982 the Defense Department unified many of its disparate military space operations through the creation of a major new division of the US Air Force, futuristically named Space Command. At the time of writing there are also plans to include the space-related activities of both the Army and the Navy within the same organization. It cannot be long before that favourite invention of science fiction writers, the United States Space Force, truly comes into existence.

Space Command has its headquarters at Peterson Air Force Base near Cheyenne Mountain in Colorado, and it also runs a number of smaller bases around the world for controlling space operations. Space Command is responsible for overall management of all Air Force space programmes which currently have an annual budget of over $10 billion – about 50 per cent more than NASA receives for civilian missions. Its duties include day-to-day control of numerous orbiting military satellites; tracking and monitoring of all orbiting spacecraft to detect possible threats to US vehicles; development and operation of new weapon systems capable of destroying hostile Soviet spacecraft; and soon Space Command will take over from NASA direct responsibility for all military Shuttle missions.

Vandenberg Air Force Base in California is the location of the military's own purpose-built Shuttle launch complex. Recently completed at a cost of some $2.5 billion, the new centre is called Slick Six, an abbreviation of its official title: Space Launch Complex 6, or SLC 6. Slick Six is the third spaceport on the planet capable of handling manned missions. It is situated on a desolate promontory of land jutting out into the Pacific Ocean and the new complex is intended to complement NASA's Kennedy Space Center in Florida in two important ways. First, Vandenberg provides a high-security facility for classified military Shuttle operations that are otherwise difficult to keep secret in the relatively public environment of Cape Canaveral. Second, Shuttles can be launched due south from Slick Six and the first piece of land they encouter is Antarctica. This makes it safe to launch Shuttles into polar orbit with plenty of ocean available for splashdown of the recoverable solid rocket boosters and the large expendable external tank.

Vandenberg Air Force Base showing Slick Six Shuttle Launch pad with *Enterprise*, NASA's non-operational Shuttle that is being used to test the new launch facilities.

The ability to send manned missions into polar orbit, the first time in the history of spaceflight that this has been possible, is of considerable military significance. In low polar orbit a spacecraft will circle the Earth in approximately 90 minutes, crossing in turn both the North and the South Poles. On each circuit of the globe the planet turns about 20 degrees on its axis, and so the orbiting spacecraft will eventually overfly every part of the planet – including, of course, all Russian and American territory. Both the United States and the Soviet Union have long exploited such orbits using the sort of unmanned reconnaissance satellites described just below. But now that Slick Six allows manned polar operations on the Shuttle, the United States will enjoy two important advantages. In the first place, it will be possible for military astronauts to carry out their own comprehensive observations of foreign territory from space; and secondly, the Shuttle will be able to deploy and then revisit unmanned surveillance satellites in polar orbit, which should enhance reliability and allow for extended spacecraft lifetimes.

The Titan expendable launch vehicle that the Pentagon view as an essential back-up for the vulnerable and unreliable Shuttle.

It would, however, be wrong to conclude that the Defense Department is unreservedly delighted with the Shuttle – it most certainly is not! Having rescued the project from cancellation more than once during the 1970s, the military are now having to face some unpleasant realities. One difficulty has been a number of irritating delays caused by problems with the Orbiter's heat-resistant tiles and, to a lesser extent, with the Shuttle main engines that have forced cancellation of several military Shuttle missions. Far more serious of course, is the loss of the entire *Challenger* Orbiter. With chilling foresight a US Air Force spokesman predicted exactly this eventuality in 1983: 'Any unforeseen difficulties with the Shuttle, which could conceivably include a catastrophic loss of a vehicle and its crew, will certainly cause the entire system to be grounded for an indeterminate length of time . . . Sole reliance on four Shuttle Orbiters represents an unacceptable national security risk.'

To overcome what it now sees as a dangerous over-reliance on one launch system, the Air Force recently decided to commission ten commercial expendable boosters to supplement Shuttle missions. These boosters are similar to the Titan T-34D workhorse that has carried numerous large American military payloads into orbit to date. They offer the added advantage that a launch can take place at very short notice: something that is not possible with the Shuttle but which is often essential for intelligence-gathering operations.

## Spies in the sky

Unmanned military reconnaissance satellites are probably the most numerous and yet at the same time the least publicized category of space missions. The Defense Department is understandably reluctant to release any information about its capabilities in this area and, in fact, the very existence of American spy satellites was officially denied until

President Carter made a reference to 'national technical means of verification' in a speech about the SALT treaty in 1978. What little is known has slipped out in the form of leaks in the aerospace press, particularly in the magazine *Aviation Week & Space Technology* which published a list of reconnaissance satellite launches in its edition of 17 December 1984. The magazine has also printed photographs of Soviet shipyards, showing the construction of enormous aircraft carriers, which were almost certainly taken from space.

According to *Aviation Week & Space Technology* the United States has flown more than 55 reconnaissance satellites, all launched on Titan boosters, between 1970 and 1984. The majority of these spacecraft carried relatively simple high-resolution cameras, with each mission directed at a particular intelligence target. But the Air Force also has at its disposal more sophisticated devices such as the 'Big Bird' broad-coverage reconnaissance satellites that can return six separate re-entry pods with exposed film during a service life of up to 180 days. 'Big Bird' vehicles are probably able to drop down to very low altitudes for brief periods to obtain close-up shots of areas of particular interest. (It is interesting to note that the year when the greatest number of 'Big Bird' satellites were launched – 1973 – was the year of the Yom Kippur War.)

In addition to the large 'Big Bird' spacecraft, the United States Air Force has also launched a lesser number of KH-11 digital transmission spy satellites which can send back their images by radio. The lifetime of these massive vehicles, which like 'Big Bird' are believed to weigh about 13,500 kg (30,000 lb), is considerably extended since mission duration is no longer constrained by limited supplies of film. The Air Force is believed to be working on a new version of the KH-11 to be

According to Jane's Defense Weekly, this image is a computer-enhanced U.S. satellite picture of a Soviet nuclear-powered aircraft carrier under construction at Nikolaiev Shipyard 444 on the Black Sea. Classified state-of-the-art spy satellite imagery is doubtless even better.

launched by the Shuttle once Vandenberg becomes operational. It is highly likely that polar Shuttles, flying out of Slick Six, will visit orbiting KH-11 satellites to refuel and maintain them.

Less is known about Russian spy satellites, although it is an easy matter to identify numerous reconnaissance missions from their characteristic low-altitude polar orbit and their relatively short flight duration. The Soviet Union has launched an average of about 30 such missions a year which constitute approximately one-third of all Russian space operations. In addition, it would appear that cosmonauts on board the Salyut space station have occasionally carried out photographic work for intelligence purposes. In fact, Salyut 3 and Salyut 5 were placed into an unusually low orbit – 241 km (150 miles) as opposed to the more usual 322 km (200 miles) – which may well have been intended to facilitate reconnaissance photography.

But just how detailed are the pictures from state-of-the-art spy satellites operating under favourable conditions? Rough calculations based on the likely maximum size of the optics carried on board such spacecraft, and the minimum altitude at which they can safely fly, suggest that it should just about be possible to read from orbit the headlines of a newspaper on the ground. In practice, adverse weather conditions and less than optimum lighting mean that performance will often be much worse than this. However, even the ability to photograph features with a resolution of about half a metre, which is undoubtedly achievable, is of enormous strategic importance. With such intelligence it is possible to count and identify aircraft on the ground, to spot the construction of new missile silos, and to monitor naval exercises and troop movements.

The future holds in store a host of new methods of surveillance from space. Already a sophisticated radar imaging system has been tested on the Shuttle and, according to *Aviation Week & Space Technology*, it will soon be incorporated into a new class of military surveillance spacecraft to be launched from Vandenberg by the Shuttle. The new radar system, which probably uses technology common to the Venus Radar Mapper probe, will yield photograph-like images of the ground at night and even through thick cloud. Such a capability would be particularly useful for keeping track of tanks and other armoured Warsaw Pact vehicles.

Ocean surveillance satellites are another important area of development and it is even possible that some of today's Soviet spacecraft may be able to detect the passage of deeply submerged nuclear submarines by the barely perceptible wake they create on the surface. (Cosmos 954 which, embarrassingly for the Russians, crashed into northern Canada in 1978, was probably on just such a mission. It carried a small nuclear reactor on board to meet the considerable power demands of an ocean surveillance radar.) Another significant project, this time American, is the classified Teal Ruby programme in which a staring array of infrared sensors on an orbiting satellite is used automatically to track the movements of aircraft in the atmosphere below. Technology derived from the Teal Ruby project will undoubtedly be installed in future military spacecraft.

Surveillance of the enemy is, however, only one of the services provided by orbiting military spacecraft. $C^3$ – Command, Control and Communication – the key to military effectiveness, depends today more than ever on the existence of satellites. Satellites allow commanders to communicate with units anywhere in the world, at any time of day and regardless of weather conditions. Accurate forecasting of these weather conditions, which is essential for successful planning of military operations, would be all but impossible without the benefit of instant pictures from meteorological satellites. Finally navigation satellites, such as the 18-strong American Navstar constellation, can now be used to determine position with an accuracy of just a few metres. Military users, ranging from lost foot soldiers to submarine-launched nuclear missiles emerging from the ocean, require only a compact lightweight terminal to receive and interpret the satellites' signals.

### Vulnerable assets

With so many valuable assets in orbit it is hardly surprising that both the United States and the Soviet Union have developed the technology to attack one another's satellites in time of war. ASAT systems, as anti-satellite weapons are known, pose a significant threat to the balance of military power in the world and there is considerable international pressure to bring these weapons within the scope of arms limitation talks. The present ASAT capabilities of both sides are still fairly

A U.S. Air Force F-15 fighter launches an anti-satellite missile (ASAT) that ascends to orbital height to intercept its target. Both fighter and missile are controlled by computers at Space Command in Colorado.

primitive and, in particular, key satellites in distant geostationary orbit remain secure from attack. But anti-satellite warfare is currently a hot topic with considerable investment being channelled into research and development. There can be little doubt that more sophisticated weapon systems are in the pipeline.

The United States had an early lead in ASAT technology with a variety of research projects, including crude anti-satellite nuclear-tipped missiles in the 1960s. The Soviet Union then undertook extensive development and testing of a killer satellite system that, by the late 1970s, had demonstrated on several occasions the potential to destroy a variety of strategically important low-orbit US spacecraft. The Russian ASAT system works by launching a killer satellite to intercept an orbiting target at a high closing velocity. Just before encounter, the killer satellite explodes to produce a fast-moving and widely dispersed cloud of shrapnel which destroys the target like a blast from a shotgun. The Soviet system can achieve rendezvous within only one orbit, giving little time for evasive manoeuvres, and although tests have so far been limited to relatively low altitude, there seems to be no theoretical reason why the range of the Soviet system could not be increased.

Seriously concerned by this hostile capability, the United States has now developed its own operational ASAT system which works using an entirely different principle. The American weapon consists of a special warhead carried into space by a small two-stage solid-fuel rocket, which is itself launched from a specially modified F-15 fighter. The F-15 fighter is under the command of the Space Defense Operations Center which is part of Space Command in Cheyenne Mountain. SPADOC monitors the trajectory of a target Soviet satellite and transmits to a distant F-15 fighter the co-ordinates of a point in space where the satellite will be at a known moment in the future. The pilot of the F-15 then puts his aircraft into a steep climb and, at the appropriate instant, he fires the ASAT missile which hurtles up into space to intercept the target.

The small 16 kg (35 lb) warhead of the American ASAT missile carries no explosive charge. Instead it relies on an extremely sophisticated guidance system to ensure that it collides successfully with the target satellite, using the combined kinetic energy of both vehicles to ensure their mutual destruction. During the final phase of its mission, the ASAT warhead goes into a rapid spin along the axis of its flight path to ensure stability. An array of infrared telescopes in the vehicle's nose then lock on to the heat signature of the target satellite and powerful guidance computers fire precise charges of solid propellant to make final corrections to the trajectory. If all goes according to plan, the two vehicles will meet with a closing velocity of approximately 8 km (5 miles) per second and the Soviet satellite will be reduced to small fragments.

The United States claims that it has developed its ASAT capability primarily to deter the Soviet Union from ever employing hers. Of the two systems, there is little doubt that the American weapon is more flexible and probably more reliable, but it is limited to a maximum altitude of about 1000 km (620 miles) and up to 70 per cent of potential Soviet target satellites orbit above this limit. This performance gap is unlikely to last long, however, particularly given the high level of American expenditure on research into new types of space weapon.

**Star Wars**

On 10 June 1984 a Minuteman intercontinental ballistic missile issued forth from a silo at Vandenberg Air Force Base and headed out west on a trajectory aimed at Kwajalein Atoll, 8000 km (5000 miles) away across the Pacific Ocean. Twenty minutes later on Mech Island, as the incoming warhead appeared over the horizon, another rocket blasted off and streaked into the sky. This second missile was also a Minuteman, but one that had been modified to become an experimental interceptor. It was guided by optical sensors (similar to those installed on the ASAT missile) that use long-wavelength infrared radiation to spot an enemy warhead against the cold backdrop of space.

The warhead and the interceptor met at an altitude of approximately 150 km (90 miles) with a closing velocity of more than 6000 m (nearly 20,000 ft) per second. Moments before impact the interceptor deployed an umbrella-like framework, 4 m (13 ft) in diameter, designed to increase the probability of a collision. In the resulting encounter both vehicles were utterly destroyed and debris from the Minuteman warhead was dispersed over an area of 40 $km^2$ (15 $miles^2$).

This impressive demonstration – the equivalent of stopping a bullet with a bullet – was the final test in a series known as the Homing Overlay Experiment conducted by the United States Army. HOE, like the Air Force's ASAT missile, is just a foretaste of the 'Star Wars' technology that President Reagan has proposed America should develop. In a dramatic speech delivered on 23 March 1983, the President called on America's scientists to find 'the means of rendering nuclear weapons impotent and obsolete . . . What if free people could live secure in the knowledge that their security did not rest on the threat of instant US retaliation to deter a Soviet attack – that we could intercept and destroy strategic ballistic missiles before they reached our own soil or that of our allies?'

According to President Reagan's 'Star Wars' vision, America will one day be defended by a fleet of orbiting battle stations, equipped with lasers and other devices capable of destroying Soviet nuclear missiles as they fly towards their target. The space-based network would be supplemented by a variety of weapons on the ground, creating a layered system of defence. At each level fewer and fewer Soviet warheads would penetrate, until ultimately none reaches its target. This science fiction scenario has an official name – the Strategic Defence Initiative –

and a massive budget of some $26 billion to be spent during the next few years on developing the requisite technology.

The Strategic Defence Initiative (SDI) is headed by Lieutenant General James Abrahamson, former director of NASA's Shuttle programme and of the US Air Force's successful F-16 fighter project. A considerable proportion of the money at his disposal is being spent on the development of powerful new weapons that could be used to attack Soviet ICBMs during all stages of their flight towards America. The most spectacular technology being considered is, the laser.

Lasers come in many different varieties but so far none of them look very promising as means of destroying nuclear missiles. Infrared chemical lasers have already been developed by the Pentagon with versions working on the ground at power levels of several megawatts. The trouble, however, is that they are bulky, they require massive amounts of hydrogen and fluorine fuel, and many potential targets are good reflectors of infrared radiation. So attention is moving towards shorter-wavelength lasers, particularly X-ray lasers, which would not only produce more destructive types of radiation, but which would also be much more compact and hence suitable for mounting on satellites. But here the trouble is the power source, which so far has to be a nuclear explosion! Furthermore, X-rays cannot penetrate the atmosphere below about 100 km (60 miles), which means they could not attack ICBMs during their most vulnerable boost phase.

Another weapon system being studied by the SDI office is the electromagnetic cannon or rail gun. This device uses powerful magnetic fields to accelerate small metal slugs to very high velocities. Ground tests have already achieved velocities of 5 km (3 miles) per second and the target is to achieve 100 km (60 miles) per second in space. As many as 60 projectiles per second, each weighing several grams, would be fired. In orbit the system would be powered by a nuclear reactor driving a homopolar generator that could supply large peak currents to cryogenic magnets. Analysis has shown that a space-based rail gun could deliver more energy per unit area at a range of 2000 km (1245 miles) than lasers or particle-beam weapons.

But even if such extraordinary weapons can be made to work in orbit, there are still numerous problems with President Reagan's 'Star Wars' proposal. Quite apart from the threat posed by cruise missiles and low-level bombers (which SDI ignores) there are, to say the least, some practical difficulties involved in installing an anti-missile defence system in space. The foremost of these difficulties concerns the required effectiveness of any such shield which must be very close to 100 per cent if the whole project is to be worth while. Should even an extremely small proportion of the Soviet Union's 10,000 or so strategic nuclear warheads slip through the net, then the consequences for America would be so appalling as to render the system pointless. As the protest slogan says: you can only die once.

The task of stopping 10,000 simultaneously launched nuclear warheads can no longer be likened to stopping a bullet with a bullet. Instead it is like trying to stop each pellet in a multiple shotgun blast with devices that are as yet little more than theoretical possibilities. Even assuming that enormously powerful lasers or electromagnetic rail guns can be made to work in space, and destroy fast-moving targets at a distance of many hundreds of kilometres, the following non-trivial problems remain to be overcome.

Numerous massive battle stations, maybe as many as 100, must be installed in orbit together with adequate supplies of fuel. This would require a launch capability far greater than anything now on the drawing board and the cost to the American people, over and above developing and constructing the weapons themselves, would be enormous.

In time of war, the system must be able rapidly to detect enemy launches and instruct the space weapons where to point in a timely and co-ordinated manner. This is possibly the greatest technical challenge facing the SDI office for its scientists and engineers must essentially

An artist's impression of an orbital battle station that would destroy enemy missiles by firing high-velocity projectiles using an electromagnetic cannon. Many scientists doubt that such technology could ever work with sufficient reliability to be worthwhile.

design computers that can take over the management of a nuclear battle in time of war. With 10,000 warheads potentially launched in a period of just a few minutes, the system faces a computational nightmare. It must detect and track each hostile warhead; it must tell each battle station which target to attack; it must discover which warheads have not been destroyed and instruct other battle stations to consider them; it must be aware of destruction or failure of any parts of its own system; and finally it must never launch an attack without being sure it is responding to a real missile strike. It must make all these decisions within the space of just a few seconds: certainly less time than it would take to contact the President.

If just 100 orbiting battle stations are to destroy up to 10,000 warheads then each must be capable of attacking several missiles during their few minutes of flight. Individual weapons, which will probably weigh many tons, must be retargeted and stabilized for further firing with just seconds between salvoes. They must point with an accuracy of just a few seconds of arc and they must slew to new co-ordinates many degrees away. The Space Telescope, which represents the state of the art in pointing technology, takes up to 30 minutes to move between different parts of the sky and it then requires a further 15 minutes to stabilize. During that time, all Soviet ICBMs would have completed their trajectories!

Even if these challenges are overcome, it is foolish to suppose that the Soviet Union will sit idly by while a means to negate her nuclear forces is established. Battle stations must themselves be protected against attack by a host of anti-satellite weapons including co-orbiting killer satellites, nuclear-tipped missiles, ground-based or space-based high-energy lasers, and homing interceptors. Recognizing this danger, the SDI office has been considering 'pop-up' systems which would be launched from silos or submarines at short notice in time of war. But apart from increasing the cost still further, this approach means that the valuable opportunity to attack missiles during the boost phase would be lost.

These obstacles to a 'Star Wars' defence system are so severe that it is now widely accepted that a leakproof shield is impossible to achieve within the foreseeable future. Indeed, the former head of the Pentagon's laser weapon programme has now admitted, 'we will not be seeking to develop a single system which can intercept and flawlessly defend against all missiles and all attacks.' Instead $26 billion is being spent on the SDI to discover the more general potential of new space-based weapon systems. SDI will also put pressure on the Soviet Union: politically at disarmament negotiations and economically in forcing the Russians to develop equivalent technologies. President Reagan may have sold SDI to the American people as a solution to the threat of nuclear weapons, but in reality it looks like just another cynical move in the arms race.

CHAPTER EIGHT

# Russia In Space

*The Soviet Union is the sea coast of the Universe.*
Konstantin Tsiolkovsky, Soviet space pioneer

Since Yuri Gagarin became the first human being to orbit the Earth on 12 April 1961, Soviet cosmonauts have logged more than 10 man-years in space. This impressive total, which is almost three times the corresponding American figure, reflects a will to support a sustained human presence in space that has frequently been lacking in the West. Year in, year out, Soviet cosmonauts have blasted off from their Tyuratam spaceport to stay in orbit for weeks and sometimes for months at a time. Gradually and unspectacularly, the Soviet Union has established a bridgehead in orbit that in some respects now represents a more significant achievement than NASA's technologically more advanced Shuttle programme.

Appreciation of Soviet accomplishments is not helped, of course, by the fact that we tend to hear so little news about them in the West. Newspaper and television coverage of space concentrates on NASA's activities while Russian missions are mentioned briefly, if at all. But this apparent censorship is not entirely due to Western chauvinsm. In contrast to NASA's ever-eager publicity machine, the Soviet Union releases only the bare minimum of information concerning its space operations, and then only so long as everything is going according to plan. Furthermore, the Russian space programme largely lacks the sort of science fiction spectaculars – like untethered space walks and satellite rescues – that make good news copy.

Six cosmonauts, five Russians and an Indian, crowd the Salyut-7 space station.

**A typical three-month extract from the busy log of unmanned Soviet space activity.**

| *Date* | *Mission* | *Orbit (km)* | *Inclined* | *Probable purpose* |
|---|---|---|---|---|
| 04/07/84 | Cosmos 1581 | 40,165×614 | 62.8° | Early warning |
| 05/07/84 | (Meteor 2) | 974×954 | 82.5° | Advanced weather observation |
| 19/07/84 | Cosmos 1582 | 274×249 | 82.3° | Earth resources film return |
| 24/07/84 | Cosmos 1583 | 412×354 | 72.9° | Military reconnaissance |
| ??/07/84 | Cosmos 1584 | 268×193 | 82.4° | Earth resources imaging |
| 31/07/84 | Cosmos 1585 | 324×181 | 64.8° | Imaging reconnaissance |
| 02/08/84 | Cosmos 1586 | 40,165×614 | 62.8° | Missile early warning |
| 02/08/84 | (Gorizont) | Geosynchronous orbit | – | Telecommunications |
| 06/08/84 | Cosmos 1587 | 394×209 | 72.9° | Photo reconnaissance |
| 08/08/84 | Cosmos 1588 | 457×438 | 65.0° | Electronic ocean surveillance |
| 08/08/84 | Cosmos 1589 | 1,523×1,500 | 82.6° | Earth geodetic research |
| 10/08/84 | (Molniya 1) | 40,772×479 | 62.7° | Military communications |
| 16/08/84 | Cosmos 1590 | 293×221 | 82.4° | Earth resources |
| 24/08/84 | (Ekran) | Geosynchronous orbit | – | Television relay |
| 30/08/84 | Cosmos 1591 | 300×220 | 82.3° | Earth resources film return |
| 04/09/84 | Cosmos 1592 | 380×202 | 72.9° | Military reconnaissance |
| 04/09/84 | Cosmos 1593-5 | 19,141 circular | 64.7° | 3×Glonass navigational spacecraft |
| 07/09/84 | Cosmos 1596 | 39,342×613 | 62.8° | Missile early warning |
| 13/09/84 | Cosmos 1597 | 272×219 | 82.3° | Earth resources film return |
| 13/09/84 | Cosmos 1598 | 1,029×987 | 83.0° | Navigation |
| 25/09/84 | Cosmos 1599 | 275×179 | 67.2° | Reconnaissance |
| 27/09/84 | Cosmos 1600 | 404×215 | 70.0° | Military reconnaissance |
| 27/09/84 | Cosmos 1601 | 521×477 | 65.8° | Military technology |
| 28/09/84 | Cosmos 1602 | 680×648 | 82.5° | Ferret military signals interception |
| 28/09/84 | Cosmos 1603 | Manoeuvrable | – | Military development vehicle |

Nevertheless, in terms of the sheer scale of its operations, the Soviet Union is the number one spacefaring nation on the planet. In recent years its annual launch totals have consistently approached and often surpassed the 100 mark, which is more than four times the level of American space activity. Its aggregate payload now exceeds 1 million kg (2.2 million lb or 1000 metric tons) in orbit. A glance at a breakdown of Soviet launches for a recent three-month period reveals a mixture of military, commercial and scientific missions essentially similar to the picture in the West.

## Space Station Salyut

Inevitably it is the Soviet Union's manned operations that attract the most interest and attention. The Russian programme involves a family of different spacecraft centred around the Salyut space station. To date there have been seven orbital versions of Salyut; the first, Salyut 1, was launched as long ago as April 1971, marking the tenth anniversary of Gagarin's historic flight. The current model, Salyut 7, was first occupied in May 1982 and it has been used intensively ever since. On more than one occasion Salyut 7 has required major in-orbit repairs and this fact, together with the station's busy duty cycle, implies that a replacement may soon be launched.

The vital statistics of the Soviet space station have been officially released. Salyut 7 is shaped rather like a tapering cylinder, 15 m (49 ft) long and just over 4 m (13 ft) in diameter at its widest point. The station weighs approximately 19,000 kg (42,000 lb) and it has three large steerable solar panels which can generate around 3.5 kilowatts of electrical power – by comparison, the unladen Shuttle weighs 64,000 kg (140,000 lb) and its fuel cells can generate a constant 12 kilowatts of power. An important feature of Salyut is its twin docking adaptors, one at each end of the vehicle. These ports allow combinations of other spacecraft to be attached, which is crucial to the station's design.

Inside Salyut 7, cosmonauts have 90 m$^3$ (3190 ft$^3$) of space in which to live, sleep, eat, wash and exercise during orbital tours that can last up to six months or more. The station contains a single room-like compartment about 8 m (26 ft) long and 3 m (10 ft) wide, plus two small airlock chambers at either end to serve the docking ports. There are numerous tanks to store oxygen, water, wastes and rocket fuel, filters to purify the air and radios to keep in touch with the ground. Unlike astronauts on the Shuttle, cosmonauts on board Salyut maintain only intermittent radio contact with mission control. Communications are only possible for brief periods when the orbiting spacecraft is within range of ground stations in the Soviet Union.

Salyut is equipped with several small rocket motors for manoeuvring in space. These thrusters are essential to keep the station stable, but they also have another important function. Orbiting at a height of between 200 and 350 km (125 and 220 miles), residual traces of the Earth's atmosphere gradually cause the spacecraft to lose speed and drop

BELOW LEFT A Soviet manned Soyuz spacecraft begins its journey to orbit from the Baikonur cosmodrome.

BELOW RIGHT The Salyut 7 space station as viewed from a visiting Soyuz ferry as it approaches to dock. A second Progress/Soyuz vehicle can be seen joined to the right of the main station module which has three large solar panels.

towards the ground. Salyut fires its rocket engines every few months to restore this lost altitude. Interestingly, the boosters that carry crew members up to Salyut can only reach the station with a full complement of three cosmonauts if Salyut's orbit has decayed towards the lower end of its range. Otherwise only two cosmonauts can make the trip.

The normal residence pattern involves a mixture of long-stay and brief-visit crew members. Long-stay cosmonauts, numbering either two or three, occupy the station for between five and eight months at a time. During their extended tour of duty they are hosts to groups or visiting cosmonauts who remain on board for about a week. Salyut has a maximum capacity of six crew members, so that at any given time there could be up to three long-stay and three brief-visit cosmonauts. So far, no long-stay crew has ever directly replaced another on orbit.

While occupied, Salyut is always linked to at least one other spacecraft. Cosmonauts arrive and depart in small 7000 kg (15,000 lb) Soyuz-T vehicles which can accommodate up to three crew members. These spacecraft are launched on *semyorka* boosters, the workhorse of the Soviet space programme which has been used successfully on more than 1000 occasions. So far it is the only Russian rocket to be man-rated. Return to Earth takes place in a small subsection of the Soyuz vehicle called the descent module, and a hard landing is usually made in the Kazakhstan area of the USSR. The entire technology, apart from improvements to control electronics and propellant management, is fundamentally the same as that used by Yuri Gagarin.

**Progress and expansion**

As well as manned Soyuz transports, the Salyut station can dock with two other kinds of visiting spacecraft. The more usual type are unmanned Progress supply vehicles that regularly bring fuel, air, water, stores, equipment, even mail, up from Earth. The Progress spacecraft, which are derived from the Soyuz design, have been highly automated and they can achieve docking and fuel transfers to an unattended Salyut. This ability to replenish expendables, and regularly to supply spare parts and new equipment, is an essential step forward in achieving a permanent Soviet presence in orbit.

The final member of the spacecraft family that makes up the Salyut system has, at the time of writing, visited the station only once. Cosmos 1443 was an unmanned vehicle that docked with Salyut 7 in March 1983. It represents a new design of module quite unlike the Soyuz or Progress vehicles. If anything, it more closely resembles Salyut itself. Approximately 13 m (43 ft) long and 3 m (10ft) across, the new expansion module weighs an impressive 20,000 kg (44,000 lb) including all the supplies with which it is filled. It has two solar panels to augment the electrical power available from Salyut and it also has its own propulsion system to assist with orbital manoeuvres. Like the Progress vehicles, the Cosmos 1443 class module is capable of resupplying

Salyut with fuel, oxygen and water, but unlike Progress, the new unit increases the habitual volume of the station by an additional 50 $m^3$ (1765 $ft^3$). It also includes a sizeable unmanned re-entry capsule for returning several hundred kilograms of cargo to Earth.

When it was launched the new module looked for all the world like a permanent extension to Salyut 7. Western observers predicted that when the re-entry capsule detached, an additional docking port would be revealed – restoring the Salyut complex to its normal twin-port configuration. But when, in August 1983, the cosmonauts duly employed the re-entry capsule to send some cargo home, the whole of Cosmos 1443 separated from the Soyuz-Salyut complex. Nine days later the capsule itself parted from the main vehicle and returned safely to Earth. Then, in September 1983, controllers on the ground fired Cosmos 1443's rocket engines and sent it plunging to destruction in the atmosphere.

Despite this rather surprising turn of events there can be little doubt that the new module does represent an important addition to Salyut flight hardware. Spacecraft designers and flight controllers in interviews with Russian journalists have talked freely about the importance of Cosmos 1443 class expansion modules for the Salyut system. James Oberg, a respected Western authority on the Soviet space programme who works at NASA's Johnson Space Center, quotes chief cosmonaut Vladimar Shatalov in his recent book *The New Race For Space*:

> This modular execution makes it possible to extend the facilities of orbital scientific complexes. A module can be special purpose: an astronomical observatory, a geophysics laboratory, a greenhouse for biological research and experiments,. a technological workshop . . . In short there are many different versions. Eventually I can imagine the future of Soviet astronauts – this is space complexes composed of several units. There will be everything on them for normal life, work, rest and sports. The units will contain research equipment, laboratories and workshops. I think that some of the units will remain aloft for some time in an autonomous flight without communication with the station, according to a set programme. They will be docked to the basic station for repair, exchange of equipment and supplies of medical preparations, extrapure materials, etc.

### Cosmonauts at work

Vladimir Shatalov's list of applications for the new class of Salyut expansion models echoes the range of activities currently performed on board the station. (It also duplicates many of the suggested applications for NASA's Space Station.) During their record-breaking 237-day orbital tour in 1984, cosmonauts Leonid Kizim, Vladimir Solovyev and Oleg Atkov carried out literally hundreds of different experiments in numerous branches of science. They also undertook limited production runs of valuable made-in-space products, including purification of pharmaceuticals using electrophoresis equipment similar to the Shuttle-based EOS system.

So far the most successful industrial operations conducted on Salyut have employed large electrical furnaces to grow high-quality crystals. The Korund furnace, which weighs about 150 kg (330 lb), was delivered to Salyut in an unmanned Progress freighter. It uses a microcomputer accurately to control the temperature inside the furnace to within half a degree C. It can run for up to three days unattended and it can process samples weighing several kilograms, heating them to a temperature as high as 1270°C. The Korund furnace has mainly been used to produce crystals for use in the electronics industry, including gallium arsenide, cadmium selenide, and indium antimonide.

Experiments involving electrophoresis and high-temperature crystal growth demand that Salyut should be kept as still as possible during production runs to minimize disruptive acceleration forces. This means that another of the cosmonauts' principal occupations – astronomy and Earth observation – must be scheduled at other times. Both activities require Salyut's telescopes and observation windows to be pointed in specific directions and this involves frequently firing the station's thrusters.

For Earth resources work an MKF-6M camera array, manufactured by the famous Karl Zeiss Jena factory in East Germany, is employed. There is apparently a healthy demand for such Salyut photographs and more than 600 Soviet organizations have made use of the images to assist with mapping, land management, pest control and geological surveys. Even the Ministry of Fishing has become involved, training cosmonauts to detect subtle differences in ocean conditions that betray the likely presence of fish shoals. Apparently Salyut 7 saved Russian fishing fleets more than 20 million roubles during 1983.

In the field of astronomy, the primary instrument on board Salyut 7 is a 500 kg (1100 lb) X-ray spectrometer housed in a bulky 'dustbin' that protrudes into the crew's living quarters. The telescope has been used to make observations of major X-ray sources such as Cygnus X-1, which is one of the most promising black hole candidates in the sky. Cosmonauts have also used a French astronomical camera, supplied as part of the cooperation in space activities between the two nations.

Indeed, the political significance of Salyut should not be underestimated. Aside from the international propaganda value of maintaining a vigorous space programme, the Soviet Union has enhanced its reputation abroad through a scheme known as Intercosmos, whereby representatives of 11 other nations have now flown in space. The countries involved are: Czechoslovakia, Poland, East Germany, Bulgaria, Hungary, Vietnam, Cuba, Mongolia, Romania, France and India. James Oberg has observed wryly that cosmonauts from the six main Soviet bloc nations flew in strict Russian Cyrillic alphabetic order – which does not say much for the scientific or technical considerations underlying their selection! In all cases the guest cosmonauts have joined missions paying brief visits to long-duration Salyut crews.

Yet for all the politics and propaganda, and for all the lip service to industrial and economic benefit, undoubtedly the greatest achievement of the Salyut programme has been the sheer number of man-hours in space now accumulated by Soviet cosmonauts. Ten men have stayed on Salyut for missions lasting 150 days or more and one of them, Valery Ryumin, holds the world endurance record with a cumulative total of 361 days in orbit. It can hardly have been pleasant at times, isolated from home and cramped into a cluttered and noisy machine not much larger than a caravan. But the Soviet achievement is important because it shows that human beings can adapt to long periods of weightlessness and because it paves the way for some really exciting missions in the future.

Soviet mission control near Moscow.

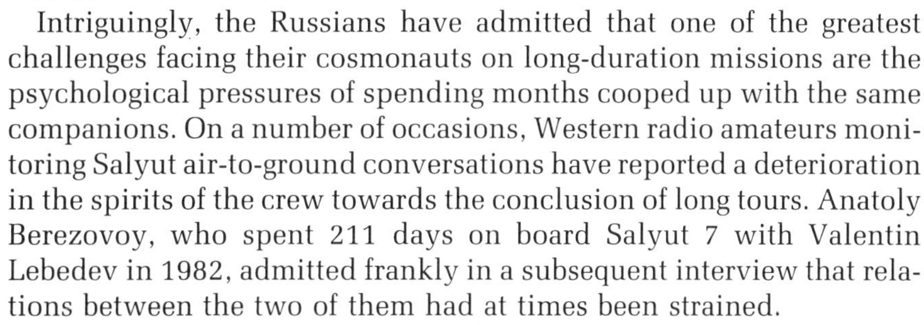

Intriguingly, the Russians have admitted that one of the greatest challenges facing their cosmonauts on long-duration missions are the psychological pressures of spending months cooped up with the same companions. On a number of occasions, Western radio amateurs monitoring Salyut air-to-ground conversations have reported a deterioration in the spirits of the crew towards the conclusion of long tours. Anatoly Berezovoy, who spent 211 days on board Salyut 7 with Valentin Lebedev in 1982, admitted frankly in a subsequent interview that relations between the two of them had at times been strained.

In an attempt to improve the psychological climate on board Salyut, Soviet doctors have encouraged the cosmonauts to take part in regular two-way television conferences with their families back on Earth. Mission control has also learned that it is frequently better not to assign specific tasks to each crew member, but instead to leave the cosmonauts themselves to distribute the daily list of duties as they see fit. Precisely the same conclusion was reached when relations between astronauts and mission control deteriorated badly during the final American Skylab mission in 1974.

In another parallel with Skylab, Soviet cosmonauts on board Salyut have demonstrated the value of having 'men in the loop' to intervene when things go wrong. Just like Skylab, Salyut 7 developed trouble with its solar panels, seriously restricting the amount of electricity available to power the station. Apparently things got so bad in the autumn of 1983 that cosmonauts Vladimir Lyakhov and Aleksandr Aleksandrov found themselves inhabiting a cold and damp spacecraft with moisture condensing on all exposed surfaces. Quite apart from being uncomfortable, this situation created a major electrical hazard and seriously threatened the continued viability of the station.

When rumours of their predicament first reached the West, it occasioned a spate of alarmist reports which suggested the cosmonauts were marooned in space. These rumours were fuelled when a rescue mission carrying a specially trained repair crew blew up on the pad in September 1983. Fortunately cosmonauts Vladimir Titov and Gennady Streaklov were carried to safety by their launch escape rocket, but not before

LEFT Cosmonaut Valery Ryumin (right) holds the endurance record for the longest cumulative period in space: he has been away from Earth for a total of almost a year.

their booster had been engulfed in flames. Apparently they sustained no serious injuries and they were reportedly well enough to consume a glass of vodka before being taken off to hospital for a medical check!

This accident left the unhappy crew up in Salyut 7 no better off than before, and now they were forced to undertake the necessary repairs themselves. During two separate sorties from their spacecraft they attached new solar panels (previously delivered on an unmanned Progress freighter) on to the side of the existing arrays. The value of having cosmonauts on board to undertake repairs was underlined again the following year. In April 1984, during their record 237-day mission, Leonid Kizim and Vladimir Solovyev left the confines of the station to replace a leaking exterior propellant valve that was preventing normal refuelling operations.

Further evidence of a growing Soviet extra-vehicular capability came in October 1984 when Svetlana Savitskaya (the second woman in space participating in her second mission) successfully experimented with special metal cutting tools while floating outside Salyut 7. Savitskaya's excursion, which made her the first woman to walk in space, occurred just shortly before Kathryn Sullivan performed a similar exercise from the Shuttle – thus achieving yet another space first for the Soviet Union.

BELOW Svetlana Savitskaya becomes the first woman to walk in space. Salyut 7's folding solar arrays are visible behind the cosmonaut to the right of frame.

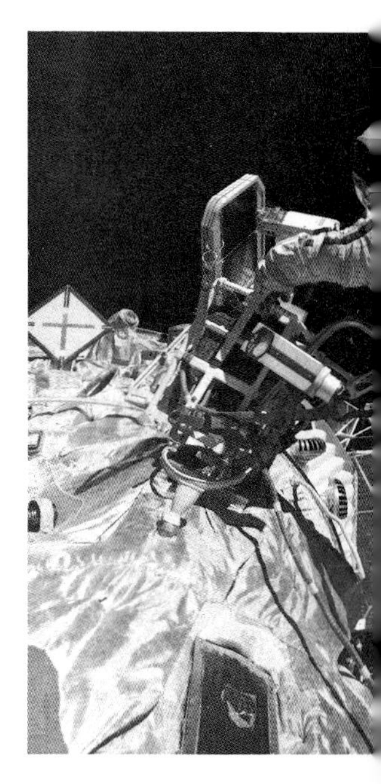

**Things to come**

Predicting the future of the Soviet space programme is a dangerous pursuit, but there do seem to be some significant and reasonably likely developments on the horizon. The most anticipated event is the unveiling of a Russian version of the Shuttle, which is believed to be at an advanced stage of construction. On four occasions in recent years, unmanned scale models have been sent into orbit and have shortly thereafter returned safely to Earth. So far all the mini-shuttles that have flown have landed in the sea: at first in the Indian Ocean where photographs of the spacecraft were obtained by an Australian reconnaissance aircraft, and latterly in the Black Sea, which indicates a greater ability to

RIGHT Soviet cosmonaut Valery Kubasov and Hungarian cosmonaut Bertalan Farkas follow tradition and chalk their names on the side of the landing capsule after returning safely to Earth.

control the vehicle. There are also reports of free-flight atmospheric tests on a full-scale prototype being conducted from a Bison bomber mothership.

The Soviet shuttle is known to have many similarities to NASA's craft and, needless to say, there have been plenty of American claims that it is merely a copy. Like the original Shuttle, it employs heat-resistant tiles for protection against the fierce temperatures of re-entry, and its dimensions are believed to be almost identical. According to a report in *Aerospace Daily*, apparently based on American spy satellite photographs, the Russian shuttle is 33.2 m (109 ft) long and 23.2 m (76 ft) wide. The corresponding measurements for NASA's Orbiter are 37.2 m (122 ft) by 23.8 m (78 ft). One important difference, however, is that the Soviet vehicle does not have any rocket engines on board. Instead the main propulsion plant is attached to the base of an external fuel tank which, like NASA's ET, is discarded several minutes after launch. If it is true that the Russian system lacks re-usable engines, then it is probably because Soviet cryogenic propulsion technology is less advanced.

A Soviet shuttle is not the only major development predicted shortly to occur. Since the early 1980s there have been persistent American intelligence reports that the Russians are building a huge superbooster: a rocket even more powerful than the mighty Saturn V that carried Apollo to the Moon. In August of 1984 the magazine *Aviation Week & Space Technology* claimed that the United States Air Force spy satellites had photographed 'two large space boosters now undergoing advanced checkout at the Tyuratam launch site'. According to this report, one of these was the core of a superbooster. When fully assembled, it will tower more than 100 m (330 ft) and it will be able to carry a payload weighing 150,000 kg (330,000 lb) into orbit. That is five times the capacity of the American Shuttle.

If such reports of the superbooster are well founded – and it is likely the Russians will have tested their new rocket by the time these words appear in print – then the Soviet Union will have gained the ability to

The Salyut 7 space station (above) is thought to be near the end of its design life by the time this book is published a much larger Soviet complex may be in orbit.

place really enormous structures into orbit. With a single launch the Russians could orbit a payload equivalent to NASA's Space Station, whereas using the Shuttle it will take the best part of a year to assemble the structure from separately delivered subsections. But what are the Russians planning to do with their new rocket? The United States Department of Defense claims that the superbooster is primarily intended for launching vast laser battle stations, and it has used this assertion to help justify its own 'Star Wars' Strategic Defence Initiative.

To be less pessimistic, there are two civilian applications for the new Soviet rocket that could transform the Soviet space programme. The first possibility is that the superbooster will be used to launch a vastly expanded version of Salyut, establishing a large permanently manned space station in orbit. Soviet space personnel have frequently referred to such a 'Cosmograd' – a city in the sky. With a likely capacity of about 12 people, 'city' may be an exaggeration, but such a project would nonetheless overshadow NASA's eight-person Space Station which, in any case, will not become operational until well into the 1990s.

The second potential civilian application for the new Soviet superbooster is even more dramatic. During their 237 days they spent orbiting the Earth on board Salyut 7 in 1984, cosmonauts Kizim, Solovyev and Atkov covered a distance of approximately 155 million km (95 million miles). Both the duration and the distance are roughly what would be involved in a round-trip manned mission to Mars. There is no firm evidence that the Soviet Union *is* planning such an expedition in the immediate future. But Russian cosmonauts and senior figures from the Soviet space programme have repeatedly stated in interviews that a red flag on Mars is definitely a long-term objective. The superbooster will, for the first time, provide the means to assemble a large expeditionary spacecraft in Earth orbit. Having lost the race to land a man on the Moon during the 1960s, the red planet must offer a tempting target for Soviet space planners looking to the 1990s.

**CHAPTER NINE**

# Visions

*'Tis strange – but true; for truth is always strange; stranger than fiction.*
From *Don Juan*, Lord Byron

In his novel *Space*, James Michener records that President Roosevelt once summoned a team of America's leading scientists to the White House. The President asked them to outline for him the major scientific developments that lay ahead. 'After three days of intense speculation these men, whose job it was to anticipate the future and who commanded more keys to the future than any other group, failed to predict atomic power, radar, rockets, jet aircraft, computers, xerography, and penicillin: all of which were to burst upon the world within the next few years. They knew about the exploratory research, of course, but they could not believe it would produce functional products so soon.'

There is little reason to believe that our ability to predict scientific developments has much improved since Roosevelt's time – particularly in space where political as well as technological considerations determine the shape of things to come. So far this book has been cautious in describing the future in space: most of the missions discussed in earlier chapters have already been allocated funds and are firmly scheduled to occur within the next few years. But beyond the mid-1990s the picture becomes much less clear. So, following boldly in the footsteps of

Roosevelt's brains trust, this final chapter explores some of the more ambitious, exciting, though uncertain developments on the horizon.

**Return to the Moon**

Twelve men, all Americans, have walked on the Moon. Gene Cernan, the last to leave, climbed the ladder of his Apollo 17 Lunar Module shortly before Christmas 1972: 'I take Man's last steps from the surface for some time to come – but, we believe, not too long into the future.' More than a dozen years later, America still has no firm plans to return to the Moon, but at least there are now signs that Cernan's hopes may be realized within his own lifetime. In November 1984, the National Academy of Sciences organized a conference of 300 scientists, engineers and astronauts who met to discuss the idea of establishing a permanently manned scientific station on the Moon. This conference was sponsored by NASA with the aim of encouraging long-term planning for a lunar base, and there was a growing sense of optimism among the participants.

Two factors now suggest that America will return to the Moon sooner rather than later. First, there is the decision to develop Space Station which, once established in orbit, will markedly reduce the cost of lunar missions. Re-usable orbital transfer vehicles (OTVs), initially intended to deliver and service geostationary satellites, will provide an economical means of reaching the Moon. In fact, paradoxically, it actually takes less energy for an OTV to travel from the Space Station to lunar orbit, than it does for the same spacecraft to rendezvous with a geostationary spacecraft. Thus the most expensive component of a lunar base, namely a two-way transportation system, will already be substantially in place.

The second reason why interest is reviving in a return to the Moon stems from the perception that America needs a long-term space goal that transcends the essentially practical functions of the Space Station. Indeed, around the time President Reagan made his commitment to that project, it was widely rumoured that he would simultaneously endorse the creation of a lunar base. In the event, the $180 billion Federal budget deficit soon dashed any hopes of such an ambitious and expensive programme getting off the ground, but tacit acknowledgment was made that the Space Station is not, in itself, a sufficiently visionary long-term objective for NASA.

Presidential science adviser George Keyworth confirmed this view in his comments at the NAS Moonbase conference in November 1984. Not normally noted for the warmth of his commitment to manned space operations, Keyworth was surprisingly bullish about a return to the Moon: 'The lunar base is one of the more obvious of the bold and exciting goals that we can reach using the Space Station as a doorway . . . Remember that much of the momentum of our space programme after Apollo was lost because we treated the Moon landing as an end unto itself. This time we should know enough to define and update our

The last men on the Moon.

goals in space in broad terms related to our future, not just in individual projects.'

So what is the point of returning to the Moon and establishing a permanently manned base there? Well first, despite six Apollo landings and three automated Soviet sample-return missions, lunar science still lacks a great deal of data. In fact, Mars is today more thoroughly mapped than the Moon! Sample sites are confined to a narrow central area and many aspects of lunar geology remain mysterious. The Moon's ancient rocks, largely preserved from the destructive effects of erosion, provide a valuable record of events in the solar system going back billions of years. Planetary science would benefit enormously from a comprehensive lunar survey and the Moon still holds many secrets.

A lunar base, however, offers other scientific benefits, beyond simply an improved understanding of the Moon itself. With almost no atmosphere, the Moon is the ideal place to locate a huge optical telescope: tracking, stabilization and thermal control would all be easier than in the weightlessness of orbit, and astronauts would always be on hand to undertake maintenance and repair. The idea of a farside radio observatory has also been proposed. The Moon's bulk would shield antennae from man-made interference generated by transmitters on Earth and a large automated listening post could be constructed to detect radio signals from other civilizations.

A Moonbase could also have practical applications. The single most important restriction on human expansion into space is the enormously high cost of launching materials from Earth. The Moon, with no atmosphere and only one-sixth of terrestrial gravity, is a much easier place to leave, as testified by the small size of the lunar module in which Apollo astronauts began their journey home. The Moon is potentially a rich source of readily accessible raw materials which could be used to develop a technological infrastructure in Earth orbit and beyond.

Buzz Aldrin, who together with Neil Armstrong made the first lunar landing on 20 July 1969, has recently become an enthusiastic advocate of mining the Moon for raw materials. He argues that production plants could be set up to supply oxygen, rocket fuel and construction metals from lunar ores. Power for these processes would come from solar energy and the final products could be delivered into orbit using mass drivers (which are described below). Oxygen would be obtained by solar heating of the mineral ilmenite, and silane rocket fuel could be manufactured from silicon-bearing rocks. If water were one day found (perhaps in deep polar craters) then hydrogen might also be produced.

Lunar mining operations may eventually become a reality, but first the Moonbase itself must be built. All the concepts advanced at the NAS conference involve use of the Space Station as a staging post, where lunar missions could be assembled, and where orbital transfer vehicles could be refuelled. According to a typical proposal, the lunar base would be established in two phases. First there would be a period of

One of NASA's more optimistic artist's impressions depicts a lunar base early in the next century. Oxide-rich minerals are being mined and processed, providing lunar oxygen to fuel a busy space economy.

manned reconnaissance with the aim of locating a favourable site for the permanent facility. In the second phase, OTVs would transport components of the base to the Moon and construction work would begin.

The reconnaissance phase alone would require 12 Shuttle launches to deliver the necessary hardware to the Space Station. The first two lunar sorties would not be manned. Instead, automated OTVs with special descent stages would deposit a pair of rover vehicles on the lunar surface and install a spare landing module in lunar orbit to back up subsequent manned operations. Then four astronauts would travel to the Moon and spend about 30 days on the surface, exploring potential sites for the final base using the lunar rovers. Having established a suitable location, they would blast off in the ascent stage of their landing module and rendezvous with an unmanned OTV in lunar orbit for the trip back home. Probably the returning OTVs would plunge briefly into the earth's atmosphere to shed excess speed before docking with the Space Station.

The assembly phase would follow, involving no fewer than nine lunar sorties. The first five missions, which would be unmanned, would deliver key components of the base – such as living quarters, a laboratory and a power plant – into lunar orbit. Then a construction team of seven astronauts would arrive in two contingents. They would begin assembling the base with the help of the rover vehicles left behind after the reconnaissance phase. When the work was done, two final missions would be used to exchange the construction crew for a science crew, and scientific operations could begin.

Cost is, of course, the crucial question. According to most estimates, constructing a lunar base will require $50-90 billion, putting the project in the same league as the Apollo programme which consumed around $75 billion in 1985 dollars. At current levels of funding, NASA would be forced to spread the cost of a Moonbase over a long period of time, perhaps as much as 25 years. This would bring peak-year expenditure within NASA's current means and annual funding would be on a similar scale to Shuttle and Space Station development. But such a protracted schedule means it will be well into the next century before a manned lunar base can be established. Modesty is necessary, however, if funds are to be won in the current economic climate, and at least there could be a return to the Moon within the lifetime of the original Apollo astronauts.

## Mining the asteroids

The Moon may be our nearest permanent neighbour in space, but from time to time the Earth receives transient visitors that pass closer to home. Indeed, sometimes visiting asteroids come rather too close and crash into our planet with devastating results. The shape of many coastlines has been altered by the impact of large meteorites, although the cleansing influence of wind, rain and sea prevents us from discerning the wider effects of this bombardment today. It is a sobering thought to realize that, on average, at least one large asteroid crashes into our planet every few ten thousand years.

However, according to some far-sighted scientists and engineers, the existence of Earth-approaching asteroids could turn out to be a blessing rather than a curse. They suggest that such asteroids, measuring just a few kilometres across, could be mined for a range of valuable materials that would be cheaper to deliver to Earth orbit than similar products obtained from the Moon. In particular, certain asteroids are rich in rare metals such a platinum, silver and gold, and others known to contain

In lunar orbit, a ferry is about to carry a cargo of liquid oxygen back to Earth. A small manned station is depicted to the top right of frame together with a lunar lander and an orbital oxygen store.

significant quantities of carbon and water which are both vital for a sustained human presence in space. Detailed studies have been made of manned mining expeditions to selected asteroids. Missions have been identified that would cost billions of dollars to fund, take many months to complete, but which could return to the vicinity of Earth with a profitable cargo of valuable materials.

Extraordinary as such an enterprise sounds, it is actually within the capabilities of current space technology. Dr Brian O'Leary, a scientist and former NASA astronaut, has designed a two-year asteroid mining mission that would use the existing Shuttle hardware. Seven Shuttle launches would be required to assemble the expedition craft in Earth orbit. It would have a crew of six to ten astronauts, who would live in modules derived from Spacelab or Space Station designs. These modules, together with equipment for mining the asteroid, would be powered by seven Centaur rocket engines burning liquid hydrogen. After a journey through interplanetary space lasting several months, the spacecraft would rendezvous with the target asteroid by firing a single solid-fuelled rocket, similar to the inertial upper stage currently used to place satellites in geostationary orbit.

Once at the asteroid, the crew would spend a busy few months refining the precious metals it contains. They would employ large, lightweight mirrors to focus sunlight and melt small sections of its surface. Using a technique known as zone refining, the concentration of target metals would build up until a fairly pure product precipitates. Brian O'Leary estimates that about 100 metric tons of platinum group metals, worth at least $1 billion at today's prices, could be collected in this way. In addition, the crew would also have to extract about 100 metric tons of water from the asteroid – either directly in the form of ice, or else by heating hydrated ores. This water would be electrolysed with solar electricity to produce hydrogen and oxygen, which could then be liquified and used to replenish the fuel tanks for the journey home.

Even this mission is child's play compared to some asteroid mining ventures that have been suggested. One ambitious proposal envisages the return of large 100 m (330 ft) chunks of asteroid wrapped in enormous parachute-like bags for processing back in Earth orbit! Propulsion would be provided by a device known as a mass driver, which has already been mentioned as a means to send lunar material into orbit.

The principle behind the mass driver is simple: a payload is placed in an iron bucket and accelerated along an electromagnetic track to shoot out of the far end at very high velocity. The obvious use for such a system is simply to launch whatever is placed in the bucket. But the mass driver can also function as a propulsion system that exploits the reaction force experienced by the device itself each time a canister is ejected. In the case of an asteroid, the mass driver would work like an electromagnetic rocket powered either by nuclear energy or solar elec-

The surface of Mars as seen by a camera on board one of the two robot Viking explorers that landed on the planet in 1976.

tricity. Pulverized rock would be used as an exhaust material, propelling the bulk of the asteroid back to Earth.

The orbits of several Earth-approaching asteroids are well known and a number of potential targets for mining missions have already been identified. One particular asteroid, known as 1982 DB, is especially suitable thanks to an unusually favourable encounter that will occur, appropriately enough, in the year 2001. A study at NASA's Jet Propulsion Laboratory has shown that at the time of closest approach, a single Centaur booster could haul 500 metric tons of material from 1982 DB back to Earth orbit! This opportunity may well stimulate more research into asteroid mining and further impetus is bound to come from the unmanned Near Earth Asteroid Rendezvous mission that NASA is planning for the early 1990s.

**Life on Mars**

The creation of a permanently manned lunar base, or the launch of a mining expedition to an asteroid, will certainly be exciting and important steps in the conquest of space. But neither project is likely to capture public attention in the way the first Moon landing caught the imagination of the world. The next occasion when space exploration will once again occupy the centre stage of history is almost certain to be the first manned mission to Mars. It is striking to realize that we are probably now closer in time to that event than we are to the flight of Apollo 11 two decades ago.

The United States, in deciding to develop the Space Station, has effectively taken a significant step in the direction of Mars. Manned missions beyond the orbit of the Moon will inevitably involve larger spacecraft than can sensibly be launched from Earth at a single time. The answer is to assemble interplanetary vehicles from a number of smaller components delivered separately into orbit. The Space Station, when it becomes fully operational in the mid-1990s, will provide the ideal location for such an activity to take place and it will markedly reduce the cost of a Martian expedition.

The Soviet Union has already demonstrated exactly this capability with its Salyut family of spacecraft. As described in the previous chapter, the Russians regularly employ multiple docking techniques to assemble an orbiting complex of vehicles, and there can be little doubt that their technology could readily extend to the addition of a Martian landing module and a powerful interplanetary propulsion stage. Soviet cosmonauts have spent many months away from Earth, so they are certainly well placed to endure the long flight times required for the journey to Mars. Indeed Valery Ryumin, who has logged more than 11 months on board Salyut, has publicly stated such readiness: 'If an expedition to Mars were being prepared and it should be necessary to hold a year-long stay in space as an intermediate step, I think we would be ready to agree to such work.'

Other Soviet space officials have confirmed their intention to send a manned mission to the red planet. Addressing a group of European scientists in 1980, chief Soviet space doctor Oleg Gazenko said: 'It is difficult to give an exact date for a flight to Mars. But I think the basic prerequisites for such a flight exist now . . . Whether the flight happens in 10, 15, or 20 years, I cannot say. But I believe it will be before the year 2000.' Since Gazenko spoke those words, the new superbooster has arrived on the launch pad, providing the Soviet Union with a means to deliver massive payloads to Earth orbit. Furthermore, plans have recently been announced for an unmanned Soviet Mars survey mission to take place later in the decade. Both developments are essential steps towards a future manned expedition: the overall pattern of Russian activity is certainly consistent with the aim of landing a cosmonaut on Mars before the turn of the century.

From a technical point of view, a manned Mars expedition offers some interesting challenges. At its closest approach to Earth, Mars is over 54 million km (34 million miles) away, which is more than 140 times further than the Moon. Even radio signals travelling at the speed of light will take at least six minutes to complete the round trip, rendering normal two-way conversation impossible! A spacecraft will take months to cover each leg of the journey. And Martian gravity, although only 38 per cent that of Earth, is more than twice the Moon's – which means large quantities of fuel will be needed when the time comes to return home.

On the positive side, however, Mars does have a thin but significant atmosphere. This could be used for aerobraking on arrival to achieve orbital insertion with very little expenditure of propellant. Mars also has two small moonlets, Deimos and Phobos, measuring only 13 km (8 miles) and 23 km (14 miles) across. Pictures from NASA's Viking missions show that these two moonlets are irregular lumps of rock, and it is likely that both are, in fact, captured asteroids. The pictures also show deep fissures penetrating their surface, which are believed to have been caused by outgassing of volatile substances. If this hypothesis is correct, then the expedition could rendezvous with Phobos or Deimos and extract enough water to manufacture liquid oxygen and liquid hydrogen rocket fuel for the return trip. This manoeuvre would markedly reduce the mass of the spacecraft departing Earth and a faster journey time could result.

An artist's impression of Robert Forward's laser-pushed lightsail starship.

**To the stars**

A manned expedition to Mars is already within the capability of existing technology, Given the necessary will, both superpowers could probably launch a mission before the end of the decade. But looking deeper into space and further into the future, what are the prospects for travel beyond the solar system? Mars may be 140 times further away than the Moon, but Alpha Centuri, the closest star, is 700,000 times

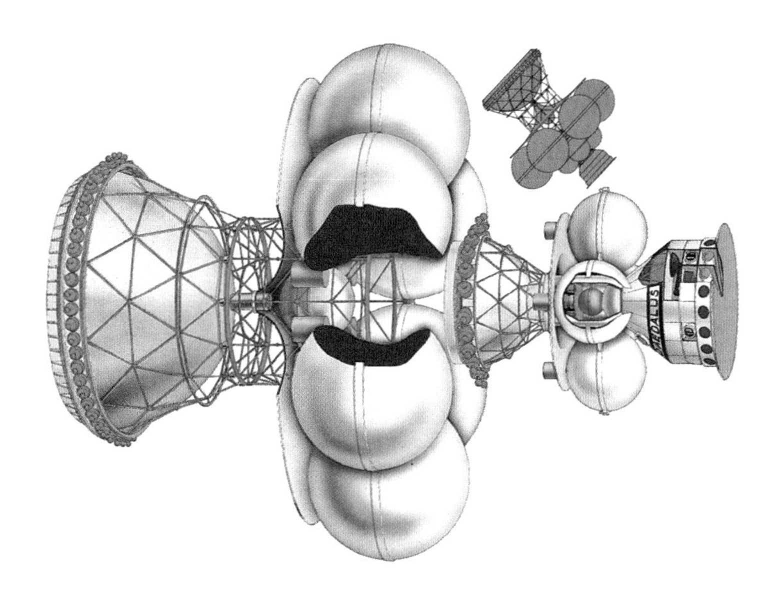

Another ambitious starship design: Project Daedalus would weigh 54,000 metric tons, require 150 Shuttle launches to deliver its components into orbit, and involve a stop-over at Jupiter to fill the fuel tanks of its fusion engine with a rare isotope of helium!

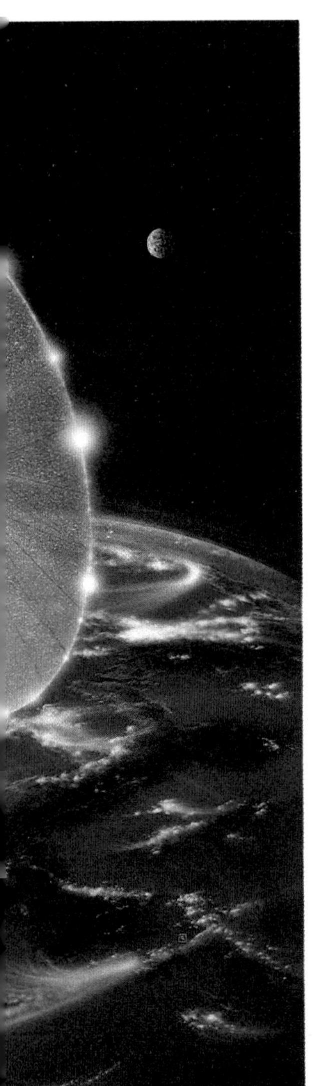

further away than Mars! At a velocity of 40,000 km/h (25,000 mph) – which is the fastest that any human being has travelled – it would take no less than 100,000 *years* to reach Alpha Centuri. Clearly some new technology is called for.

As far as we know, the only way to get around the vacuum of space is by using, in one form or another, a rocket engine: warp factors and hyperdrives are the product of fantasy not physics. Ejecting hot exhaust gas at a high velocity, a spaceship pushes itself forward by the principle of equal and opposite reaction. Chemical rockets, as used on the Shuttle for example, depend on energy released when substances like oxygen and hydrogen are combined. However, chemical rockets will never be very much more powerful than they are today, and so starship designers must search for alternative ways of producing thrust.

Nuclear energy offers some hope, promising modest improvements in the short term using fission, and major progress in the longer term when we learn to master fusion. In nuclear rockets of the first type, a fission reactor is used to heat hydrogen gas to very high temperatures. The hot gas is ejected through a conventional rocket nozzle, and the system has the advantage that all the expendable propellant is in the form of mass-efficient hydrogen. The resulting thrust is about 70 per cent greater than an equivalent chemical engine and, in addition, the nuclear system has a much greater endurance. A nuclear fission rocket has already been tested on the ground and it looks promising for use on interplanetary missions: the first manned mission to Mars may well employ such an engine. For interstellar travel, however, the improvement is insignificant and a more radical solution must be found.

Nuclear fusion will probably drive the first starship leaving Earth. Using the same source of energy as the Sun and the hydrogen bomb, a fusion rocket could cut the journey time to the nearest star from 100,000 years to just a few decades. Fusion power works by combining light elements (like hydrogen, helium and lithium) into heavier elements, releasing vast quantities of energy in the process. The problem is to persuade the light elements to fuse, which they will do only at very high temperatures and pressures. A great deal of work is currently going on around the world to harness the fusion process for use in terrestrial power stations and there is no doubt that the theory itself is sound. The technical difficulties are formidable: particularly in finding ways to confine and control a hot plasma at several million degrees Celcius.

Despite the fact that fusion technology is still many years away, a remarkably detailed design for a fusion-powered starship has already been produced by a team of scientists and engineers from the British Interplanetary Society. Known as Project Daedalus, the spacecraft would weigh some 54,000 metric tonnes fully fuelled, and it would take something like 150 Shuttle launches to deliver the necessary components into orbit. Daedalus would be unmanned and it would take approximately 50 years to arrive at its destination, Barnard's star, which is about 6 light years from Earth.

'Destination' is actually something of a misnomer, since the starship could not afford the fuel necessary to decelerate. Instead it would charge through the stellar system at close to 40,000 km (25,000 miles) per second, or approximately 13 per cent the speed of light, using a swarm of probes to make thousands of independent obervations. After an encounter with Barnard's star lasting just a few hours, the spacecraft would head out into the vastness of the galaxy. Meanwhile its precious radio signals would begin their six-year journey back to Earth, where they would be eagerly awaited by a team of scientists long since retired!

About the most charitable thing that can be said about Project Daedalus is that it does not contravene any known laws of physics! But the practical problems to be overcome are daunting. One immediate difficulty is that Daedalus' fusion system depends on a reaction between rare isotopes of hydrogen and helium. The former can be extracted from seawater reasonably cheaply, but the latter, helium 3, does not occur naturally on Earth. Instead it must be manufactured in the laboratory at a cost of several million dollars per killogram – and the starship would require about 30,000 metric tons of the stuff! Helium 3 is found in the atmosphere of Jupiter, however, so the design team propose a preliminary excursion to the outer Solar System to tank up before departing for Barnard's star!

Fusion is not the only hope for interstellar travel. Matter-antimatter annihilation is another theoretical possibility that is potentially even more powerful: it could cut interstellar journey times from decades to just years. And because matter-antimatter starships could travel close to the speed of light, relativistic time-dilation would become significant. The crew could undertake a journey lasting centuries from the point of view of people back on Earth, and yet in their own time they would return home after only a few years. This effect may sound like pure science fiction, but it undoubtedly happens, and it is not impossible that people now reading this book will one day experience such things for themselves.

MARCH-APRIL 1984 J. SPACECRAFT 187

**dtrip Interstellar Travel Using Laser-Pushed Lightsails**

Robert L. Forward*
*Hughes Research Laboratories, Malibu, California*

This paper discusses the use of solar system-based lasers to push large lightsail spacecraft over interstellar tances. The laser power system uses a 1000-km-diam. lightweight Fresnel zone lens that is capable of focusing er light over interstellar distances. A one-way interstellar flyby probe mission uses a 1000 kg (1-metric-ton), -km-diam. lightsail accelerated at 0.36 m/s² by a 65-GW laser system to 11% of the speed of light (0.11 c), ing by α Centauri after 40 years of travel. A rendezvous mission uses a 71-metric-ton, 30-km diam. payload l surrounded by a 710-metric-ton, ring-shaped decelerator sail with a 100-km outer diam. The two are unched together at an acceleration of 0.05 m/s² by a 7.2-TW laser system until they reach a coast velocity of 21 c. As they approach α Centauri, the inner payload sail detaches from the ring sail and turns its reflective rface to face the ring sail. A 26-TW laser beam from the solar system, focused by the Fresnel lens, strikes the avier ring sail, accelerating it past α Centauri. The curved surface of the ring sail focuses the laser light back to the payload sail, slowing it to a halt in the α Centauri system after a mission time of 41 years. The third ssion uses a three-stage sail for a roundtrip manned exploration of ε Eridani at 10.8 light years distance.

Robert Forward's paper in the *Journal of Spacecraft and Rockets* describing his proposal for 'Roundtrip Interstellar Travel Using Laser-Pushed Lightsails'.

There is one other proposed method of interstellar travel that deserves to be mentioned if only because it is so bold. Robert Forward, of Hughes Research Laboratories in California, has recently published a series of scientific papers with such titles as 'Roundtrip Interstellar Travel Using Laser-Pushed Lightsails'. According to Forward's proposals, an enormously powerful laser, situated in the solar system, would be used to push a lightweight spacecraft out into interstellar space. The system would exploit the same principle of light pressure that causes a laboratory radiometer to spin when exposed to the Sun. (This is due to the momentum of light photons and the principle is still essentially the same as in a rocket.) Forward has even shown how a two-stage spacecraft could be used to bounce a laser beam back and push a smaller return vehicle homewards for a roundtrip.

The most exciting thing about Robert Forward's ideas is not the detail of his calculations, but rather the daring of his imagination. At one extreme, he has proposed a project known as Starwhisp in which a 'modest' laser beam would be used to push a spacecraft weighing only a few grams to a nearby star! The spacecraft would be like a mirrored gossamer spider's web and the payload would consist simply of a few integrated circuits to make measurements at the target star and beam the data home to Earth. The elegance of this approach is breathtaking – even if the engineering is decades beyond our current capabilities.

Forward's proposals do not stop here. He has designed a roundtrip manned mission to the star Epsilon Eridani that would take 51 years to complete (although because of relativistic effects the crew would only experience an elapsed time of 46 years). The expedition would require a lightsail measuring 1000 km (620 miles) in diameter and a laser beam with a cross section of similar dimensions! The laser itself would be solar powered and Forward suggests building this 1000 km wide behemoth, which must deliver 75,000 Terrawatts of power, somewhere in the vicinity of Mercury. He adds: 'It should be noted at this point the total power output of the entire world is about 1 TW. This amount of laser power is not trivial and will require a significant commitment to build a large array of solar powered lasers in space!'

Robert Forward's magnificent laser-pushed starships will probably never be built, but we can be sure something even more extraordinary will appear in their place. Human expansion into space is as inevitable as our colonization of the six continents on Earth: the important question is how fast it will happen. Technology is no longer the limiting factor in the equation. If we wished, we could already have set up a Moonbase, launched an expedition to Mars, and established a permanently manned space station in orbit. Instead, political and economic forces restrict our progress. As Arthur C. Clarke once wrote: 'Given a sufficiently powerful motive, there seems no limit to what the human race can achieve.' The future in space will ultimately be determined, not by what we *can* do, but by what we *want* to do.

## GLOSSARY

The use of Nasaspeak has been kept to a minimum in this book, but inevitably some abbreviations have found their way into the text. Here is a list of the main ones used with a brief explanation of each.

**ASAT:** anti-satellite weapons system – such as a co-orbiting killer satellite, an air launched guided missile, or a directed beam weapon based on the ground or in space.

**CCD:** charge-coupled device – a microelectric integrated circuit that can convert a pattern of incoming light into a sequence of electrical signals that can be transmitted by radio and processed by a computer. CCDs are used in the cameras of spacecraft such as Galileo and the ST.

**EOS:** electrophoresis operations in space – a commercial programme being run jointly by NASA, McDonnell Douglas and Johnson & Johnson to manufacture high value pharmaceutical products on the Shuttle.

**ESA:** The European Space Agency – the equivalent organization to NASA in Europe. Involved in many space projects including Ariane, Spacelab, Giotto, Ulysses and Columbus.

**ET:** the external fuel tank of the Shuttle – the large orange cylinder that is discarded shortly after launch. There are long-term plans to carry a number of ETs on up to orbit as raw material for future construction projects in space.

**EVA:** extra-vehicular activity – more popularly known as a space walk. Most EVAs from the Shuttle now involve use of the MMU (see below).

**HOE:** homing overlay experiment – a US Department of Defense project to destroy incoming ballistic warheads with interceptor rockets.

**ICBM:** inter-continental ballistic missiles – the rockets that carry nuclear warheads to their destination.

**IPS:** instrument pointing system – a European-built device that allows telescopes and other instruments situated in the Shuttle's cargo bay to be pointed at astronomical targets. The IPS is an element of the Spacelab system.

**ISPM:** international solar polar mission – a two spacecraft project to investigate the Sun involving co-operation between NASA and ESA. NASA pulled out of the project leaving the single European vehicle, renamed Ulysses, to undertake the mission. Due to launch on the Shuttle in 1986.

**IUS:** inertial upper stage – a booster rocket used to transfer satellites from the Shuttle's low Earth orbit up to high altitude geostationary orbit. IUS is similar to, but larger than, PAM (see below). Both types of upper stage have failed during Shuttle satellite deployment missions.

**JPL:** the Jet Propulsion Laboratory – NASA's main facility for the design, construction and control of unmanned planetary exploration spacecraft such as Viking, Voyager and Galileo.

**JSC:** Johnson Spaceflight Center – an important NASA facility located just south of Houston in Texas. Shuttle mission control is situated on this site and JSC is also closely involved with the development of the Space Station.

**KH-11:** a type of large, low-altitude, polar orbiting spy satellite operated by the US Department of Defense.

**KSC:** Kennedy Space Center – NASA's vast launch complex to Cape Canaveral in Florida. Every American manned mission to date has left from here although KSC will soon be facing competition from Vandenberg Air Force Base in California (see SLC-6 below).

**LGO:** lunar geoscience orbiter – an unmanned mission to survey the Moon that NASA is planning for the 1990s.

**MMU:** manned manoeuvring unit – the 'Buck Rogers' backpack that Shuttle astronauts use to jet around outside the Orbiter. The MMU is particularly useful in rescuing and repairing failed satellites.

**NAS:** the National Academy of Sciences – an organization of American scientists who have recently arranged a number of important conferences devoted to the long-term future in space.

**NASA:** the National Aeronautics and Space Administration – the United States' agency responsible for civilian space operations.

**NEAR:** near-Earth asteroid rendezvous – a mission being planned by NASA for the 1990s that would involve landing an unmanned spacecraft on the surface of a visiting asteroid and possibly returning a sample to Earth.

**OMS:** orbital manoeuvring system – the two relatively small rocket engines that the Shuttle uses during the last stages of ascent, to change orbit while in space, and finally to slow itself down for re-entry.

**OMV:** orbital manoeuvring vehicle – an unmanned space tug that NASA is planning to build for use with the Space Station.

**OTV:** orbital transfer vehicle – a manned space ferry that NASA also intends for use with the Space Station. The OTV would allow astronauts to visit geostationary orbit to service communications satellites there, and even one day to return to the Moon.

**PAM:** payload assist module – a booster rocket used to transfer satellites from the Shuttle's low Earth orbit up to high altitude geostationary orbit. PAM is similar to, but smaller than, IUS (see above). Both types of upper stage have failed during Shuttle satellite deployment missions.

**QSO:** quasi-stellar object – mysterious astronomical phenomena that appear as enormously powerful radio sources at great distances. Their investigation, particularly by the Space Telescope, may help to reveal how the Universe was formed.

**SDI:** strategic defence initiative – popularly known as 'Star Wars'. The SDI is a massive military research programme announced by President Reagan and aimed at finding ways of stopping ballistic missiles and their warheads in flight. Weapons being studied include orbital laser battle stations and electromagnetic cannon.

**SLC-6:** space launch complex six – often called 'Slick Six'. SLC-6 is the Shuttle launch facility at Vandenberg Air Force Base in California that is used to send Shuttles into polar orbit for (mainly) military purposes.

**SPADOC:** Space Defense Operations Center – part of the US Air Force Space Command in Cheyenne Mountain, Colorado. SPADOC controls a wide range of military space activity including America's ASAT capability (see above).

**SRB:** solid rocket boosters – the two re-usable, solid-fuel rocket motors that are used to launch the Shuttle. The SRB's are released from the Shuttle about 130 seconds after lift-off. They fall into the ocean to be collected by special recovery ships and refurbished up to 20 times.

**SSME:** Space Shuttle main engines – the three large throttleable rocket engines that carry the Shuttle into orbit. The SSMEs burn liquid hydrogen and liquid oxygen from the external tank (see **ET** above).

**ST:** the Hubble Space Telescope – a large unmanned astronomical satellite to be launched by the Shuttle in 1986. ST will be the largest scientific spacecraft even placed in orbit. It will be able to see about seven times deeper into the Universe than any ground-based observatory.

**TDRS:** tracking and data relay satellite – a series of sophisticated communications spacecraft in geostationary orbit that NASA uses for radio links with the Shuttle, Space Telescope and many other orbiting vehicles.

## FURTHER READING

The American magazine **Aviation Week & Space Technology** is essential reading for anyone seriously interested in spaceflight. Anything of significance is likely to be reported here sooner rather than later. Annual subscription is US $53 for air mail delivery, the publishers are McGraw Hill, and their address is PO Box 1505, Neptune, New Jersey 07753, USA.

**The Illustrated Encyclopaedia of Space Technology,** principal author Kenneth Gatland, is a beautifully produced general reference work published by Salamander Books, London 1981. Although now slightly out of date, this comprehensive book covers all aspects of manned and unmanned spaceflight and includes many useful illustrations, diagrams and tables.

**The History of Manned Space Flight** by David Baker, New Cavendish Books, London 1981, is a monumental work that includes everything you could possibly want to know about the first 25 years of man in space.

A noteworthy source for space science stories is the American astronomical magazine **Sky & Telescope** which is published monthly by Sky Publishing Corporation, 49 Bay State Road, Cambridge, MA 02238-1290, USA. The edition of August 1983 includes an authoritative article about Project Galileo and the edition of April 1985 contains three useful articles about the Hubble Space Telescope.

Another excellent article on ST is 'The Space Telescope' by John Bahcall and Lyman Spitzer in **Scientific American**, July 1982.

**Project Space Station** by Brian O'Leary was published by Stackpole Books, Harrisburg, Pennsylvania, in 1983 shortly before President Reagan made his commitment to NASA's orbiting base. Although his book is now overtaken by events, O'Leary gives an enthusiastic account of what the Space Station will be like and argues convincingly that the project is worthwhile.

For an intelligent analysis of industrial opportunities in space see 'Space commercialization: how soon the payoffs?' by John M. Logsdon in **Futures** magazine, pages 71-78, February 1984. The 25 June 1985 edition of **Aviation Week & Space Technology** is entirely devoted to this subject and also now an entire journal, **Commercial Space**.

**The New Race For Space,** by James E. Oberg, Stackpole Books, Harrisburg, Pennsylvania, 1984, is an up-to-date and fairly reliable source of information about the Soviet space programme.

**Space Resources and Space Settlements,** study director Gerard K. O'Neill, is one of many NASA publications dealing with the long-term future in space and covering such topics as a lunar base and asteroid mining. It can be obtained from the Superintendent of Documents, US Government Printing Office, Washington DC 20402, USA.

Anyone interested in building a starship is referred to 'Roundtrip Interstellar Travel Using Laser-Pushed Lightsails' by Robert L. Forward in **The Journal of Spacecraft and Rockets,** March-April 1984, pages 187-195.

## ILLUSTRATION SOURCES


Anderson/Alinari, Florence: 49 (*right*)
The Associated Press: 7, 21, 22, 45
Boeing Aerospace: 115
John Calmann & King: 43 (*left*) © the Science Museum
M. W. Carroll: Title Page
European Space Agency, Paris: 39, 49 (*left*), 50-51 (*bottom*), 51 (*top*), 87 (*left*)
Ronald Grant: 25 (From the MGM release *2001: A Space Odyssey* © 1968 Metro-Goldwyn-Mayer Inc.
The Lockheed Corporation: 70, 78 (left and right)
LTV Aerospace & Defense Co/MARS, Lincs: 103
The Mansell Collection, London: 56
Martin Marietta Aerospace: 30 (top left)
McDonnell Douglas: 90-91 (bottom)
MJL Cartographics: 11, 36, 40 (*bottom*), 61, 65, 67, 76 (*top*), 79, 91 (*top*)
NASA: Frontispiece, 6, 9, 10 (3), 12, 14-15 (4), 18, 19, 23, 26, 27 (2), 30 (2), 30 (*top right*), 30 (*bottom*), 31, 35, 38, 40 (*top*), 43 (*right*), 47 (2), 52, 54-55, 58, 59, 62-63, 69, 71 (*right*), 71 (*bottom*), 75 (*bottom*) (2), 83, 88-89 (*top*), 88-89 (*bottom*), 96, 116-117, 118, 119, 120-121
Novosti Press Agency, London: 110-11, 112-113 (*bottom*), 113 (*top*), 114
Perkin-Elmer OG Photo Ltd: 74 (*top*) (2)
Popperfoto: 21, 24 (2), 85, 97, 105, 112 (*top*)
Salamander Books: 123
Science Photo Library, London: 71 (*left*) Dr Fred Espenak, 107 (*left*)
Space Command Headquarters, US Air Force: 94, 95, 99
Space Services Inc, Houston: 87 (*right*)
Rick Sternbach: 122
Rick Sternbach and Don Dixon: 82
© Tass: 107 (*right*)
US Navy: 93
Westminster Central Reference Library, London: 84 (*left*)


## INDEX

Numbers in *italics* refer to illustration captions